Ron Creasey:
Last of the Horselads

Ron Creasey:
Last of the Horselads

WILLIAM CASTLE

Old Pond Publishing

Published by
Old Pond Publishing Ltd
Dencora Business Centre
36 White House Road
Ipswich, IP1 5LT
United Kingdom

www.oldpond.com

Cover design and diagrams by Liz Whatling
Map by John Gilkes
Illustration of horse in pole harness by Alison Wilson
Typesetting by Galleon Typesetting, Ipswich
Printed in Europe

CONTENTS

ACKNOWLEDGEMENTS

I would particularly like to thank Nancy Creasey for her support and hospitality while writing this book, and Harry Buck for reading it through and making useful suggestions. Other major contributions were made by Matt Thelwell who reformatted the recordings, Caroline Brailsford who transcribed them, and Neil Lanham who allowed me to use his DVDs of Ron as additional source material. To all of them I am very grateful.

My thanks also go to Tim and Norman Caley for answering questions, to Jolyon White and Yvette Goward for their encouragement to start this book; and to Ron's daughter Anne Hughes, Faith Tilleray, and my own family for their support. I would also like to place on record my appreciation of the Morton family, in particular Geoff Morton, for my introduction into the world of the working horse. Without them I would not even have been able to ask Ron the right questions.

Finally, I would have liked to be able to thank Ron for his story, for his enthusiasm and amazing memory, for his patience in answering hours of questions and for his company and humour, which, although he is no longer with us, has often brought a smile to my face while I have been writing.

PREFACE

Most of the content of this book comes from recordings of conversations with Ron Creasey that I made in 2006 and 2007. What stood out in those conversations was Ron's ability to recall the detail of his life as a horselad. Unlike many of his generation who worked horses in their youth, Ron worked on farms which carried on using horses until the late 1950s, and after the horses had finished on the land, his continued involvement with draught horses no doubt contributed to keeping his skills and memories alive.

Even as Ron was learning the job of a farm horseman in the 1940s, it was clear that horses were being replaced by tractors, and that the whole structure of farming was changing. So while much of Ron's tale encompasses this change, his awareness of being at the end of an era also meant that he understood the value of the old skills and methods of working. He was continually asking questions and listening to the stories of the older men, so although his own farming memories only go back to the time of the Second World War, he was well aware of what happened in earlier times.

When Ron talked of his life as a horselad, living in what we nowadays would consider as basic, even harsh conditions, he expressed no sense that he was being exploited, or that he resented his lot. Others might view the way the farm lads were employed, bound to an employer for a whole year with no payment until the year was completed, as a Dickensian form of slavery, but Ron gave no hint of having been downtrodden. Ron chose to work as a hired horselad, and like many others, he accepted the system as it was. Equally, there is no romanticisation of the working horse in Ron's story, none of the 'gentle giant' syndrome so often espoused by those who look at the past through sepia-tinted glasses.

The past is simply stated as Ron saw it, as what happened. Inevitably there is some exaggeration in the telling of any story, but everyone I have spoken to commented on Ron's remarkable memory, and on occasion I have been able to confirm details of a particular event by recounting Ron's version of the story to the people concerned. This has, with only few exceptions, been the only way I have had to corroborate any of Ron's stories. In a way that is all to the good, for Ron's story is that of a farm worker, and the events are seen from his point of view. Although much has been written about the farming of this period, it has usually been written by out-siders or by the farmers themselves; those who had the time, money and literary ability to do so. Inevitably their memories reflect their position, either as an observer of the world of farming, or as the person in charge. Those like Ron, who left school at fourteen and went to work on farms as part of a society where knowledge was transmitted by example, by rules and traditional lore, and by the telling of stories, have been generally poorly served by the written word, as it was not part of their everyday world.

As part of that oral tradition, Ron was a natural storyteller, engaging, enthusiastic and amusing, and his stories moved between time and place as his mind followed a subject between different parts of his life. In putting that spoken language into a written form, it has been necessary to use my own words to explain the context and set the scene, but as much as possible I have tried to let Ron tell his own tale, straight from the horseman's mouth.

In this account there is much information about how things were done with horses, but this book is not intended to be a manual on horsemanship. Although I hope horsemen and women will discover things of interest and use, picking and choosing snippets of informa-tion without understanding the whole system can lead to trouble. The chapter on training young horses, for instance, gives details of how the horses were broken in at that time and in that part of the country. Much of it relied on doing real work in the field, having an experienced horse working alongside the youngster, and the avail-ability of competent help; conditions that rarely pertain today. The practices of pulling waggons without a britchin or pole, so the vehicle could not be effectively steered or stopped, were also common in Ron's experience, but are not recommended.

On the other hand it would also be disappointing if no-one was to use any of the information in this book in a practical way. Although Ron appreciated that his experience was of historical interest, he also used his knowledge to help many others with their own horses. Encapsulating Ron's own philosophy of life is this quotation, which Ron particularly chose to start this book:

'A man of words and not of deeds is like a garden full of weeds'.

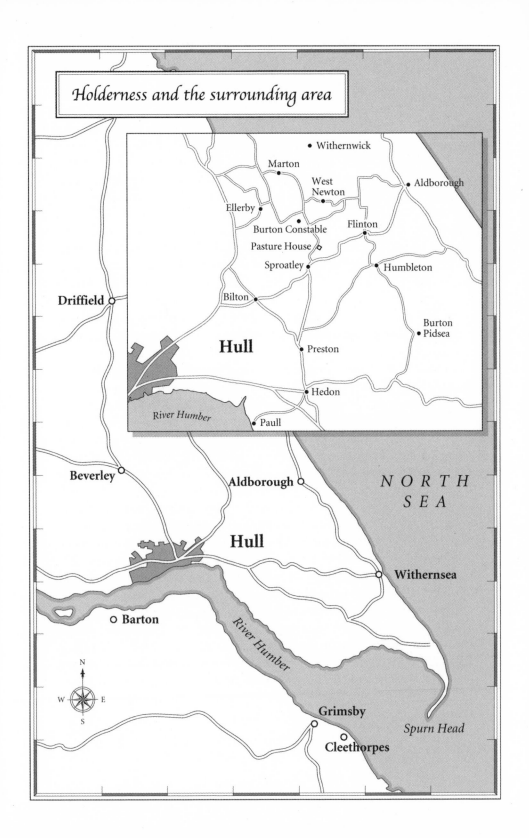

Holderness and the surrounding area

INTRODUCTION

W HEN Ron Creasey started his working life in 1943, on a farm
in the East Riding of Yorkshire, farming was going through a
period of rapid change. Some of these changes were part of a gradual
development in agriculture which accelerated because of the war,
and some changes were due to the war itself. From the government's
point of view, the priority during wartime was to produce enough
food to feed the nation. To this end, thousands of acres of grassland
were ploughed up to grow arable crops, particularly cereals and potatoes, and many farmers were forced to change the way they had been
farming, and grow different crops to meet the needs of the urban
population. All the additional cultivated acreage required both additional people and extra power to work it. Although horses provided
much of this power, at the start of the war there simply were not
enough horses, or tractors for that matter, to do the work; so tractors
were imported from America in vast numbers to cover the shortfall.

By the time war broke out in 1939, the number of farm workers
had also reached an all-time low. During the depression years farmers
were forced to cut costs in order to stay in business, and with many of
the younger farm workers joining the armed forces or going to work
elsewhere for better wages, there was a chronic shortage of labour.
To address this problem the government reinstated the Women's
Land Army, additional labour being provided by Italian and German
prisoners of war, shift workers and off-duty soldiers, many of whom
were unused to farm work. Despite the changes brought about by
the war, the general pattern of farm work continued in a manner
similar to previous years. Although tractors were increasingly being
used, particularly for the heaviest jobs, horses continued to play a
vital role, so there were still opportunities for someone like Ron
Creasey to find work as a farm horseman.

To become a horseman in the East Riding, even until the middle
of the twentieth century, did not just involve working with horses; it
meant being part of a working tradition stretching back for generations; a tradition which governed how you were employed, your
conditions of employment, as well as the way the work was carried

out. The horsemen were hired at the annual hiring fairs and were employed for a year at a time, only receiving their pay at the end of the year. During the year they had to live in the farmhouse, where their food was provided, and they only had a few hours free time at the weekends in between looking after the horses. To work with horses you also had to remain single, so most horsemen were in their teens and twenties; but no matter how old they were, they were known as horselads.

Among the horselads there was a rigid hierarchy. The horselad in charge of the stable was the waggoner, and he came first in everything; he had the best horses, it was he who took the waggon load of corn to town, and he who set the pace when working in the field. Below him came the third lad, or *Thoddy*,* then the fourth lad, then *Fiver*, and so on, depending how many horselads there were on the farm. Although the waggoner was one of the lads, he was set apart by the fact that he was responsible for the horses, the other horselads, and for making sure all the work was done properly.

Until the 1940s this system of employing horselads was still a normal part of farm life in East Yorkshire; it was what everyone expected, farmers and horselads alike. Although tractors had started to change the way in which the land was farmed, the horselads' daily routines and patterns of work were still much the same as those experienced by their fathers and grandfathers; their conditions of employment and the way they were employed having changed little during the nineteenth and early twentieth century.

In earlier times, hiring farm servants annually at a hiring fair had been a widespread custom throughout the country, but by 1900 the hiring fairs had completely disappeared from the south of England. In many northern counties, however, some farm workers were still being hired on yearly or six-monthly contracts, and in most of these counties some of the farm workers lived in; but it was only in the East and North Ridings of Yorkshire, and to a lesser extent in the West Riding and Lincolnshire, that the system of living in and being hired for a year continued. It had also been the case that both men and women were hired at the hiring fairs, whether to work on the farms or in the farm houses, but Victorian sensibilities about the two sexes mixing in public put paid to women being

---

* Italics to indicate that the word is defined in the Glossary.

hired in this fashion, so in the twentieth century only the men were hired at the fairs.

The hiring fairs were held in the week following old Martinmas day, when the horselad's yearly contract with his previous employer came to an end and he received his pay for the year. Unless he had decided to stay with the same employer, which was unusual, he would then go to one of the hiring fairs to find work for the next year. The hiring fairs were held on different days in different towns. In the marketplace, or somewhere else governed by tradition, the young men would congregate, waiting to be hired. Throughout the day, farmers from the surrounding district would come to town to seek lads to employ. One of the farmers would then approach the crowd of lads and ask one of them if he wanted hiring. If a lad recognised a farmer and knew him to be either a poor farmer or a bad employer, he might turn him down straight, but otherwise a discussion would follow. Where was the farm? How many horses did it have?

If this was a lad's first hiring fair he might be glad to be offered a job as a least lad, the lowest rung in the horsemans' hierarchy – and accept the sum offered by the farmer. If he had a few years' experience he might hold out for a bit more money or try to get a job as a third lad, or once he had spent a year or two as a third lad, he might then fancy trying for a waggoner's job, perhaps on a smaller farm.

When an agreement had been reached, the farmer gave the lad a fastening penny, or *fest*, and once this fest had been accepted, a binding contract existed between the farmer and employee, which would run until the following Martinmas. Although these agreements were verbal, traditionally everyone knew what was expected, so a breach of the rules by either party could, and sometimes did, result in a court case. In these instances, neither party could get out of the agreement unless they could prove a breach of contract. If a horselad wanted to marry, he had to wait until Martinmas; and even those lads who got a girl pregnant and had to get married during the year were still obliged to complete their contract, working and living as a horselad, as if they were still single.

For the horselads, the yearly agreement guaranteed a full year's work, and unlike in other parts of the country, they would not be laid off when work was slack or when the weather made fieldwork impossible. From the farmers' point of view, the system ensured that there was always someone available to look after the horses; someone responsible for feeding, cleaning them and mucking them out before and after the day's work. He also knew in advance how much it was going to cost.

At the end of Martinmas week, the horselads all went to their new places of work. Once on the farm, everything was governed by routine and everyone followed a strict order according to the hierarchy between the horselads. Whether coming out of the stable, riding home from the field, or helping themselves to a piece of pie, the waggoner went first, followed by the third lad, fourth lad, and so on down the line. Except on the smallest farms where the farmer acted as his own foreman, all the horselads lived in the foreman's house, or hind house as it was often called. There they received all their food and slept with the other hired lads in a bedroom that was usually directly accessible from the kitchen. The lad's food and lodging was considered to be half his wages, the other half being received in cash at the end of the year. Traditionally the wages agreed at the hiring fair covered all overtime, the time spent in the stable before and after working in the field, and the extra hours worked at harvest.

Besides the horselads, the other group of workers on the farm were the labourers. They were usually married men who lived in cottages on the farm or in the nearby villages, and because they generally were not involved with the horses, they started work later than the horselads, who had already fed and harnessed their horses before breakfast. However, there were occasions when a younger labourer might be asked to work with the horses. If more ploughs were needed, for example, one might go to plough with the horselads, but he would follow the Least Lad, even though he might well have been Waggoner himself only a year or two previously. Most of the time, however, the horselads worked the horses and the labourers did other jobs.

Once a horselad married, becoming a labourer was inevitable, as he would no longer be available at all times to look after the horses. The continual loss of horselads to marriage meant that there were always opportunities for the younger lads to move up the ladder and become a waggoner. This was part of the reason why most lads changed farms every year, but some would just fancy a change of scene while they were still young and independent. Although being Waggoner was the top of the tree for a horseman and a position which would command respect, a waggoner would not presume to give orders to a labourer. The two groups, although they often worked together, were quite distinct, both answering directly to the foreman.

The foreman was in charge of most aspects of the farm, particularly the arable side, but larger farms such as those on the Wolds would also have a shepherd and a beastman, or *bullocky*. These men usually worked independently, but at busy times they might ask the foreman to spare a lad or a labourer to help them. On the biggest farms with large numbers of sheep or cattle, they might also have a full-time assistant, and these lads, the shepherd lad and the bullocky lad, were hired by the year and lived in the hind house with the horselads. Some farms also employed a lad known as a *Tommy Owt*, who would work with the cattle, with the shepherd, or take out a team of horses as required. On smaller farms which had fewer sheep or cattle, these roles might be less defined, so in Holderness it was common for a labourer to work with the cattle, while the foreman usually looked after the sheep himself.

The foreman was the key to the smooth operation of the farm. On some of the large farms the horselads would scarcely see the farmer, and a few even lived elsewhere, leaving all the decisions to the foremen. For the horselads, the foreman was in effect the boss, so Ron Creasey would sooner say, 'I was with George Gibson,' who was the foreman, than 'I worked for John Caley,' who was the farmer. The foreman was also in charge of the stable, though usually he left the daily running to the waggoner, and his place as head of the stables is shown by the titles of the lads. Although not always the case elsewhere, on the Yorkshire Wolds and in Holderness the horselad directly under the waggoner was called the third lad. There was no 'second lad' because the second-in-command was in fact the waggoner, which reflected the foreman's position as ultimate head of the stable.

On farms throughout the country, the horseman was usually con-sidered to be the most skilled of farm workers. Not only did he know how to drive horses and how to operate all the machinery, but he was also responsible for the health of the horses and keeping them in good condition, so that they were fit enough to tackle all the work on the farm, which at certain times of the year could be arduous. It was therefore common for horsemen to be paid more than other farm workers, which reflected their level of skill and the longer hours they worked. However in the East Riding and surrounding counties that used the same system, the labourers were paid more than those

15

who worked the horses. The reason for this divergence of practice is the youth of the Yorkshire horselads, illustrated by figures from 1907.* In that year, 85% of the horselads were under twenty-five years old, and of those, nearly a third were under seventeen, a third were between seventeen and twenty, with slightly over a third being over twenty. Today, a hundred years later, approximately one-third of those lads would still be at school.

Why the farmers of the East Riding trusted their valuable horses to the care of youths, some of them straight out of school, when in other parts of the country the job was done by more mature men, can to some extent be explained by the way the whole system worked. For although the fourteen-year-old would brush his horses, muck out and go to plough with the others, he would not be allowed to feed any of the horses, and the ever-present waggoner would make sure that he was doing his work properly. Although some horse-lads became Waggoner in their late teens, many of the waggoners were in their twenties, and might already have had ten or more years' experience working with horses. Nor must it be forgotten that the foreman was in charge of the stable, and while he usually left the waggoner in charge, he would step in if necessary.

Because the system of hiring horselads was ubiquitous in the eastern half of Yorkshire, some horselads did move from one district to another, particularly onto the larger farms of the Wolds which needed many horselads; but there were also cases of lads from the Wolds moving to the surrounding areas of the Vale of Pickering and Holderness. Although the conditions of employment were the same, the working conditions in these districts did vary, so while some lads were quite happy to move into another area, others would find the change difficult and would only stay for a year before returning to more familiar territory.

The reason why working practices on farms varied in different districts is largely due to their geography. The East Riding, both geographically and agriculturally, can be divided into three, the central part being the Yorkshire Wolds, which run from north to south, veering eastwards to the coast at their northern end. At the time of the enclosures, the common land on this range of hills was divided into large arable holdings, so the farms on the Wolds

---

* Caunce, Stephen, *Amongst Farm Horses – the Horselads of East Yorkshire*, Alan Sutton Publishing, 1991.

required a large numbers of horses and a large group of horselads to work them. To the west of the Wolds is the Vale of York, where smaller mixed farms predominated, and to the east of the Wolds, bounded by the wide Humber estuary and the North Sea, is Holderness.

It was on this nearly flat, largely featureless and windswept plain that Ron Creasey started his farming career. In stark contrast to the free-draining chalk soils of the Wolds, the land in Holderness lies wet and intractable. Composed of the boulder clays left behind by the last Ice Age, interspersed with some sand and gravel deposits, and peat which has formed in depressions in the clay, most of the land has only been made workable by underground drains and the deep ditches which bound the fields. Despite being difficult to work, the soil is inherently fertile, growing good crops of wheat; but the nature of the soil made it hard to work with horses. So while the horselads on the Wolds often worked with two-furrow ploughs pulled by a team of three horses, three horses were often needed to pull a single-furrow plough in Holderness.

For the horses and for the men, farm work was hard physical work, and the only way it could be done was by long hours, skill and good organisation. Although in other parts of the country the farms were organised in different ways, the standard of work in the East Riding was as high as any, and the system undoubtedly resulted in the efficient use of both horses and men. While horses were necessary to the working of the farms, they needed to be available for work during all the normal working hours; so having the horselads on the premises to feed them in the early morning and after work in the evening was essential to the successful running of the farms.

Keeping the horselads constantly occupied also helped to keep them out of trouble most of the time; they were, after all, a group of youths and young men who were thrown together and expected to work mostly without supervision. Another regulating influence was the ethos on the farms; work was taken seriously, and everyone was expected to work hard and do a good job. The lads all had a shared interest in the farm work, and the horses in particular, and they generally took great pride in turning out the horses well, especially when taking them out on the road. Even when they were not working, much of the conversation was about work, and they identified very much with their role as a horselad and their position in the hierarchy; so much so that they would usually call each other by their title, 'Wag', 'Thoddy', or 'Fourther', rather than using each other's names.

Although the camaraderie between the lads and their desire to do a good job resulted in the horselads being largely self-policing, inevitably disputes would arise. It was then the job of the waggoner to sort things out. If a younger lad was not concentrating on his work, a sharp word or a clip around the ear would often be enough to bring him back in line. A larger lad might present a bigger problem, the standard remedy for this misbehaviour being a kick up the backside. Since the waggoner was usually older and therefore more experienced and stronger than the other lads, they would usually accept his authority, but fighting between the lads was not unknown. Most of it, however, would wait until Martinmas when the lads were free agents, and if they were still inclined to make a fight of it, they were sure to have an audience when they sought redress on hiring day. As with any system or organisation, there were inevitably individuals who were unhappy about the situation they found themselves in, but in general, the horselads were content with the hiring system because they knew that however bad a farm it was, or however rough the waggoner, it was only for a year.

Although life could be hard, and the work was hard, a horselad had the opportunity to find pleasure in his work, and through his proficiency, command respect as a useful member of rural society. The spartan living conditions were offset by large quantities of good food, probably better than many would have received at home, and working in the fresh air was definitely more healthy than working in a factory. Although a horselad's wages were lower than those of industrial workers, he had very few expenses, and because he could move between farms every year, he had a flexibility and freedom unknown to those employed in industry or apprenticed to a trade. Nonetheless, for some, working on farms was not to their liking, and many would seek other employment after a year or two. When tractors eventually came in, there were also many who were glad to leave the horses behind and sit on a tractor seat all day. But for others, like Ron Creasey, working with horses was their life.

In many ways, Ron's life as a horselad was not unusual; there were many others over the years who shared similar experiences. What is unusual is how he came to be a horselad in the first place, and that he realised the value of his experience even as a teenager. His story is that of a horselad in the final years of the hiring system, at the very end of the horse era. The heyday of the draught horse was already long past, and the hiring system had had to make adjustments to accommodate legal and social changes. So while Ron's experience cannot provide an overview of the hiring system in its entirety, or a complete picture of working practices on East Riding farms, his story does give us an understanding of the wider scene by painting a vivid picture of the life of one horselad in one corner of the Riding.

Nonetheless, this story is still very much Ron's own story, seen through his eyes, and influenced by his upbringing, his experience and his character. As a child of the depression years and as a lad starting work during the war, Ron's story reflects the social and agricultural changes of his own era as much as it describes being a farm horseman. However, because of his passion for a way of life that was fast disappearing, he was acutely aware of what was being lost, so his recollections also throw light onto the traditions and working practices of earlier times.

CHAPTER 1

Hiring Day

OLD Martinmas day, 23 November, was always an important day for a horselad in the East Riding of Yorkshire, marking the end of the yearly contract with his employer, and the day when he received his pay for the year. He then had a week's unpaid holiday, during which he had to find himself another job on another farm. For many lads this was a chance to get a better job and rise up in the horseman's hierarchy, find a place where the food was better, or to see a little bit more of the world. Only a small proportion of the horselads stayed on for another year at the same farm, so for them Martinmas week was just a chance to go home to their families, whom they may not have seen for a year if they lived far away; but most of the lads moved to a different farm every year. Some of them might already have found a new place to go to before Martinmas, but the majority went to one of the hiring fairs to get hired again for the following year.

Until the early twentieth century there were hiring fairs in some of the larger villages of the East Riding, as well as the market towns. In Holderness, the villages of Hornsea, Patrington and Hedon all held hiring fairs, but by the middle of the century – when Ron Creasey was being hired – there were only three hiring fairs in the whole of the East Riding. These were held at Driffield, Beverley and Hull.

Generally the farmers or their foremen went to the nearest hiring fair to find lads to employ, and many of the lads, perhaps especially the younger ones, would also go to the fair nearest home. There were some lads, however, who fancied working further afield, so they would travel, perhaps a twenty- or thirty-mile bicycle ride, to the hiring fair in the district where they wanted to work, rather than going to the local fair.

The hiring fairs were held on different days in each town. The Hull hiring fair, where Ron was hired, was held on the Tuesday following Martinmas Day, Martinmas Tuesday as it was called; the

Beverley hirings following on Wednesday, with Driffield on Thursday. In earlier times when there were more farm horses, the old part
of Hull where the fair was held would have been teeming with farm
lads and farmers, but by the 1940s the local people of Hull were
scarcely aware of the events taking place in the pubs and on the pave-
ments around the parish church of Holy Trinity, as old Hull was no
longer the main centre of the town.

For the farm lads, however, hiring day was a crucial day of the
year, and it was very much their day. After a year of taking orders
from the foreman and being kept in line by the waggoner, hiring day
was the day when the lads could negotiate their wages and terms of
employment as equals with the farmers. On hiring day they were
their own men; no-one could force them to go and work on a particular farm, and if they had been a third lad for a year or two, they were
free to turn down any offers of a job as a third lad and try for a
position as Waggoner.

Besides the serious issue of finding a job, the farm lads were also
out to have a good time, and most of them had money in their
pockets from the last year's work. In between talking to the bosses
and foremen, the horselads would come across other lads that they
had met on previous hiring days or at work on the farms, which led
to the development of a party atmosphere. Besides some juvenile
high spirits, which were inevitable among a group of youths and
young men, some had grievances from earlier years and scores to
settle; so on hiring day there was also some rowdy behaviour.
Indeed, since Victorian times the local newspapers had printed letters
from disgruntled townsfolk, complaining about the behaviour of
the lads during Martinmas week. This led to efforts to put an end to
the hiring fairs, with their noise, high spirits and some aggressive
behaviour; but to little avail, as the scene was much the same in the
1940s, as Ron recalls:

> We would be walking up and down, little lads, big lads, some was
> going to fight somebody else. He would come along, would ' little
> lad, and would say to one of ' waggoners, 'You kicked my arse about
> five years ago,' and this lad had now got to be big.
>
> He said, 'What are you going to do about it?' And they'd their
> jackets off. And they'd set about some of these foremans, and say,
> 'You kicked my arse.'
>
> 'What are you going to do now?'

Although some disagreements were settled in this way, part of the success of the hiring fairs was because the tables were turned during Martinmas week: the lad who last week was higher up in the hierarchy than you was now just another lad looking for a job, and the man who had been your boss was just another farmer looking for lads so that he could keep on running his farm.

The hiring fairs also acted as a forum for exchanging information, where the lads were able to tell each other about poor employers. Always they had it in the back of their minds that even if they were unfortunate enough to end up on a place that didn't suit them, they would only be there for a year and could then move on.

> If you was a good lad and you got hired as little lads, Least Lads, you could come back to the same place as Third Lad or Waggoner. It was typical of all the lads to just *stop* a year, because you knew full-well the next place you went to was all horses and the same principle of doing things, and the same food. Some were better cooks than others, but the food was just the same. And very often they'd gone round from one place to another, to and fro, to and fro, and if they were good lads, you could allus go back if you left at ' right time, twenty-third of November.

Although the number of lads being hired at the hiring fairs diminished quickly during the 1940s, when Ron first went to the hiring fair there was still quite a throng of lads looking for employment, even though, because of the war, there were more jobs than there were lads available to fill them. Visiting the hiring fair every year was a regular ritual for all the lads, even if they had already secured a job, and from 1943 Ron went every year. But the hiring fair in 1946 would become a significant day in Ron's life because, having just turned seventeen, he had handed in his notice at the farm where he had worked for the last three years, and was now looking to find himself a job working with horses full time. Compared with Ron's first visit to the hiring fair three years earlier, there were fewer lads waiting to be hired, but the time-honoured procedure was just the same.

The lads arrived about nine o'clock in the morning and gathered on the wide flagstone pavement in front of Holy Trinity Church, waiting to be approached by the farmers. There were plenty of jobs in Ron's time, which meant that the lads were not keen to accept any job offers at the start of the morning, hoping for a better offer later in the day.

We all congregated on these flags, and you just stood with your back to the church wall. Everybody was there, all the foremans and all the bosses, they were all dodging about trying to hire you. They'd all be marching up and down, and they'd look at your appearance, and they'd come up to you and say, 'We want a little lad' or 'We want a third lad' or 'We want a fourth lad', and you'd say, 'No, I don't want to come to you.'

They didn't ask you for any real references, as long as: 'Can you run?' 'Cause if you can't, it's no good coming to me,' and then they'd say, 'Can you plough without wheels?'

You'd say, 'I never have done.'

'Well, can you push a big barrow load of muck?'

'Yis,' you'd say.

'Well, you'll be alright, you'll be able to plough without wheels.' And that was how you were engaged.

Foreman was in Hull and you're walking up and down, 'Now me lad, do you want hiring?' And if you was interested, you'd say, 'Well, what do you want?'

And they would tell you what they wanted and what they didn't want. Some of 'em would ask you where you was last year, but a lot of them knew you slightly by sight, and I was with somebody who George Gibson knew, and he come along and said, 'Do you want hiring?'

He was looking for a *fowth lad* and there was no wet days – which meant you didn't work in the rain if it was at all possible. I said I would come. So I went to John Caley's of Flinton; Carr Farm, Flinton.

And if you were engaged, then they give you a fastening penny, and once you took that fest, you were their man, and you couldn't get hired by anybody else, and if you did, you had to send your fest back. It was only a matter of half a crown or five shillings, but it was binding by court, and if you talked to the men in previous years to me, they took 'em to court for not going to the place when they'd received their fest. But things were starting to relax by ' time we were lads; they wasn't going to take you to court for half a crown, although labouring men were only earning three pounds something for a fifty-hour week.

After a lad was hired, either on the pavement or in one of the nearby pubs, the rest of the day was spent socialising. The older waggoners often congregated in the pubs, while seventeen-year-old lads like Ron spent some of their time and money on the funfair, and otherwise wandered around Hull in groups, sometimes making their presence felt as they linked hands and marched up the length of Whitefriargate, 'playing hell up'.

When they met lads they had worked with in previous years, talk would inevitably turn to the different farms where they had worked, and to the new farms where they were going. If they found that they were going to be working on farms in the same district, they would often arrange to meet up and travel at least some of the way together. In Ron's case he discovered that he would be spending more than just the short bicycle ride to Flinton in such company.

> On our rounds I run across another lad who I'd been with at Porter's two years previous, and I said, 'Now then, have you got a place?'
> He said, 'I'm going as Thod Lad to George Gibson's at Flinton.'
> I said, 'Hell, I've just got hired, I'm going as Fowth Lad!'
> 'Oh,' he said, 'we shall be together again.'

In between the socialising, there were other things to be done. Traditionally, Martinmas week was the only holiday the horselads had. Having been paid for the year, this was the time to buy a new set of clothes and new boots for the forthcoming year, and many of the shops stayed open later to cater for the increased trade.

The lads also had to arrange to get their belongings to the next farm. Although Ron only had a suitcase, many of the farm lads had a big wooden box, four or five feet long, and sixteen or eighteen inches wide and deep, in which to keep their possessions. These boxes usually had a drawer inside, where the lad might keep his watch, and they were often passed down from one generation to the next, or from an older brother who was getting married to a younger brother who was being hired for the first time. At Martinmas, the boxes were collected from the farms by the local carrier and taken to the lads' homes, and then on hiring day they were all taken into Hull. The carriers' vehicles, which in Ron's time were mostly lorries, but still included some horse-drawn vehicles, lined the sides of the road near the statue of King Billy, over the road from Holy Trinity Church. The boxes were then transferred from the carrier serving the district where the lad came from, to another carrier in whose area the new farm was located, so when the lad went to his new place of employment, his box was already there. The lads then had the rest of the week free, and even though they probably had already been hired, many would visit the other hiring fairs at Beverley and Driffield, which served the farms on the Wolds. Then on the following Sunday everyone went to their new places, thus completing the change-over between farms that had started nearly two months earlier.

In the East Riding it was not only the horselads who were employed on a yearly basis. The labourers were also employed from one Martinmas to the next, but unlike the horselads, they were paid weekly. Being largely married men, the labourers were more settled and did not move quite as often as the horselads, but it was still common for them to move between farms. Because changing jobs meant moving out of their cottages, the married labourers were given more time to find a new job, so for them the hiring process started around the time of Hull fair, three weeks earlier than that of the horselads.

> Hull fair week was a big thing in this area. It was the third biggest fair in England, and they used to run trains to it from Wales; and especially if Hull City was playing at home and Hull Kingston Rovers was at home, you can imagine the train loads that would come in. So Hull fair week, which is about ninth of October, they went round the married men and asked them if they'd stop. Because if they didn't stop, they'd time to find a house by ' time of twenty-third of November, when they'd take another job on. And if they didn't ask them, they didn't want them.

The process of hiring the horselads followed a similar pattern, each being asked about their intentions, starting with the waggoner.

> Starting on first of November, we were usually sowing wheat, he'd come round would Foreman, and we'd all be harrowing, and he'd go to Waggoner first.
>
> 'Now then, do you stop again?' Then Thod Lad, 'Do you stop again?' And Fowth Lad. Now if he didn't want any, he wouldn't ask, he'd just walk past you.
>
> 'Do you stop again?' Of course, when you're little you either say yes or no.
>
> 'Yes.'
>
> 'Oh, alright.' And maybe you was undecided, but whatever we

said the first asking, nobody took any notice. Then the same day the following week he came again.

'Now then, do you stop again?' What was said to each other you didn't know, because you were all well out of earshot of each other.

Of course, when you went into stable at dinnertime, 'Has he asked you to stop again?'

'Aye.' Well he had, but some on 'em he hadn't, and they knew they weren't wanted. So then the third week, he'd come again.

'Now then, do you stop again?' And it was getting a bit serious now, because it was getting near twenty-third of November, and this time he'd probably have Boss with him. Or Boss would come on his own and say, 'Now then, Foreman tells me you aren't stopping again.'

'No.'

'Why not?'

'We don't get enough to eat.'

He'd say, 'Oh! Well Foreman's a good feller. His wife isn't maybe much, but Foreman's a good feller, so I can't do much about that', and you'd iron your little grievances out, and then if you stopped, you stopped.

The same process was being repeated on all the farms, and because they knew that their own horselads were likely to move, all the farmers and foremen had been looking over the hedges, keeping an eye open for other good lads to hire, so often a horselad had already agreed to go and work on a neighbouring farm before Martinmas day.

Then ' twenty-third ' November come, dinnertime. Then you went in for your money. You went to Boss's house and you stood outside. Waggoner went in to ' front room, and there was money all at one side and stamp cards at the other, and then you could start and tell them what you thought about 'em. Nobody was there, just you and him, and they could tell you what they thought about you, an' all! Wag went in first. Well, if Wag wasn't stopping, he'd ask Third Lad, if he was good enough lad, to stop as Waggoner. And then if he was stopping as Waggoner, you could stop as Third Lad and then they'd get another Fowth Lad. So you got your money, all in pound notes, and maybe the odd white fiver, and you was gone.

Before leaving the farm, each lad collected his horse brasses and any other possessions he had from the stable, and then had to make

While most horselads were hired at a hiring fair, some found work through newspaper advertisements. The wording of the adverts shows the variety of job descriptions and the overlap between jobs on some farms.

Situations Vacant columns from the *Hull* and *Yorkshire Times*, 14 November 1936

Mr Ward's advert for a Waggoner/ Foreman probably meant that the farmer was active on the farm and acted as his own foreman, but the waggoner would take his place at times. To 'meat' lads means he had to feed them, in other words he was the hind.

This did not mean that the applicant would always be working with horses. The term 'married man' implied work as a labourer, but he would also be expected to work the horses when needed.

Tractor drivers were still rare in 1936, (the Caleys bought their first tractor in 1936) hence this farmer's openness to employ someone without experience.

WANTED, experienced Threshing Machine Driver.—Clark, Sherburn-in-Elmet. P16

WANTED, Wagoner to board 3 or 4 men; no small children.—Owen, Thirkleby, Malton.

WANTED, Wagoner Foreman to meat lads; good stacker.—Ward, North Dalton.

WANTED, Two Farm Labourers and Day Boy.—M. Cooper, Owmby, Grasby, Lincoln.

WANTED, Single Man, or Boy, for milking; live in.—Foxton, Benningholme, Skirlaugh.

WANTED, Single Farm Labourer; board found.—Garratt, Foreman, Wootton, Ulceby, Lincs.

WANTED, Married Man to go with horses; cottage.—Ellerington, Eske Manor, Beverley. 14

WANTED, Two Hedging Men; winter's work; lodgings available.—Jackson, Coates-by-Stow.

WANTED, Bread Baker and Roundsman; married.—G. Sellars, East Kirkby, Spilsby, Lincs. P

WANTED, Single Beastman, good milker and feeder.—R. Garness, Walk Farm, Little Weighton. P13

WANTED, Married Man, small farm; cottage, garden; knowledge of stock.—Cautley, Owstwick. 13

SHEPHERD-LABOURER wanted, house on farm; one with working lad preferred.—Hookham, Northorpe.

WANTED, Single Cowman, young, active; dry hand milker; Martinmas.—Write P 844, "Times," Hull. H

MARRIED Labourer wanted immediately; all farm work.—Fussey, Burnham Beeches, Barton-on-Humber.

WANTED, at once, Young Married Labourer; good house.—Golland, Santon, Appleby, Scunthorpe.

WANTED at Martinmas, experienced Single Shepherd.—W. D. Sellers, Low Caythorpe, Rudston, Driffield.

WANTED, Man, or strong Youth, as waggoner; also Youth, about 16, able to milk.—Wastling, Lelley.

WANTED, at Martinmas, single Beastman; good milker.—Apply F. Wood, Brandesburton Grange, Driffield. P13

WANTED, at Martinmas, good, all-round Man; able to drive tractor, or willing to learn.—J. O'Gram, Harswell, York.

WANTED, Reliable Milkman; thoroughly understands work; good house.—Spilman, Brickhills, Broughton, Brigg.

WANTED, Farm Labourer, must be good stockman; cottage and garden.—F. Brown, The Laurels, Flatgate, Howden. P

WANTED, Married Man, to milk, good cottage; also Strong Youth to milk, and go with horses.—Wright, Elstronwick. P13

Situations Vacant columns from the *Hull* and *Yorkshire Times*,

21 November 1936

WANTED, a Single Beastman; good milker.—Hornby, Manor Farm, Gox-hill.

WANTED, single Beastman, age 18 to 20.—Apply G. A. Atkinson, Wassand, Hull. P

WANTED, a single Beastman; good wage to good milker.—Write P1165, "Mail," Hull. P

WANTED, Married Labourer, good horseman; house found.—A. Dent, North Cave. 23

WANTED, married Man, able to milk.—Apply H. Dobson, Burstwick-road, Hedon. P20

WANTED, single Beastman, also 2 good Horse Lads.—Porter, The Grange, West Newton. P23

WANTED, single Beastman for Wawne Hill.—Apply F. Farnaby, Southfield, Wawne.

WANTED, Single Beastman, to milk and go with horses.—Stamford, Withernwick Grange. 24

WANTED, Wagoner Foreman to meat lads; good stacker.—Ward, North Dalton.

WANTED, good all-round Farm Hand; cottage provided.—Bew, Garton, Aldbrough. P24

WANTED, Hind; wife to manage dairy and poultry.—J. T. Marshall, Garton, Holderness. P

WANTED, experienced Farm Foreman, to board lads.—R. Nutt, Derwent House, Aughton, York. P

WANTED, Tommyought; also Youth, 16 or 17, to go with horses.—Briggs, Kirkella Grange. 23

COWMAN, single, Wanted; good wages; references required.—J. Hardy, South Kelsey, Lincoln. P

WANTED, single Beatsman; must be good milker and stock man.—Garbutt, Routh, Beverley. 23

WANTED, several single Horsemen and strong Horse Lads.—H. Caley, West Newton, Hull. P

WANTED, Married Wagoner to Board men.—Apply Campbell, Skirlington, Skipsea, Driffield.

WANTED, Wagoner and young Horse Lad.—H. F. Abram, North Farm, Shiptonthorpe, York.

WANTED, Single Wagoner, Third Lad and Beastman.—R. Garness, Walk Farm, Little Weighton. P

SHEPHERD-LABOURER wanted, house on farm; one with working lad preferred.—Hookham, Northorpe.

WANTED, young Wagoner, Third Lad, and Single Beastman. — Fewston, Brandesburton Barff, Driffield.

WANTED, Wagoner Foreman to meat lads; good stacker. Also single Beastman and Lad.—Ward, North Dalton.

WANTED, three Single Horsemen, live in; good wages given.—P. Nicholls, Island Farm, Broomfleet, East Yorks.

WANTED, Wagoner to carry; also Third Lad, 17 or 18, to feed horses.—Wastling, Walkington Grange, Beverley.

WANTED, Single Man as Tractor Driver; also Single Man to assist beastman.—Apply F. W. Gooder, Bracken, Driffield.

LABOURER Wanted; able to do all classes of farm work; with working son or girl

Because this advert specifies 'a good horseman', the successful applicant for this job would probably be expected to break in young horses. His house was part of his wages.

This single beastman, or Bullocky lad, would live in with the horselads, feed the bullocks, milk the boss's cow, and work with horses.

The Tommyought, or Tommyowt, lived in with the horselads, but would do any type of farm work.

This advert from H Caley, John and Norman's father, for several horselads shows they were advertising for workers on more than one farm, as Old Farm, West Newton needed three horselads for nine horses.

Mr Ward's advert shows that in addition to the Waggoner/Foreman advertised the previous week, they were now sure that the Beastman and Lad were not going to stay for another year, so advertised their positions too.

the first payment out of his wages, for the year's washing. Although the lads lived in the foreman's house, the foreman's wife had enough work to do keeping the house and feeding the lads, without doing the lads' washing as well, so in the week after Martinmas the lads had had to make alternative arrangements.

> Our washing consisted of a shirt and your socks. That was for the week, and then you'd a Sunday shirt and your Sunday socks, and in those days you had your collars and your collar studs. Foreman's wife wasn't going to do your washing, so you'd say to a labourer, 'Does your wife do washing?' And some on 'em said yes, and they'd do your washing each week. You'd only two lots of everything, they'd do one and you got the other back, and they used to charge us five pounds for the year, something in that region, and you didn't pay 'em 'til you got paid at Mart'mas.

After settling their washing bills, the lads returned to their homes for a week of rest and social visits with friends and family, before going to the new farms.

> It was usually about sixth of December when we got to our places, 'cause we got paid on twenty-third of November, whenever that fell, and we'd the remainder of that week off. Then the following Tuesday we went and got hired again, and then we'd the remainder of that week [as well]. So if Mart'mas day fell on a Monday you could have a fortnight off, but if it fell on a Friday, you'd only have a week.
>
> Now up to the 1930s when they were hired, they were hired for a sum of money for the year which covered all their overtime, doing their horses, the farm work and everything, and if they said they'd got twenty pound for the year, that was twenty pound they were going to get: they weren't going to get anything above or under. Whereas when we got hired they said, 'Would you like a sum every week?' Because you'd no money at all, and I think I said I wanted ten shillings, and so every week at Saturday dinnertime there was ten shillings on me plate. That's what you had throughout the year, unless you wanted more money, and then you had a sub, but they allus made sure you never subbed more than the money you earned.

Giving the lads a small amount of money every week was a change from the traditional way of being paid at the end of the year and only being allowed to sub some of that money occasionally during the year. There had also been other changes by the time Ron started work. From 1938 all workers became entitled to a week's paid holiday, which for the horselads was in addition to Martinmas week; and the introduction of the National Wages Board in 1924 fixed the wages of agricultural workers. In practice this meant there was a floor to any wage negotiation between the horselads and the farmer, so younger lads would usually accept the wage offered, while a waggoner might haggle a bit if he thought his capabilities warranted a higher wage.

These changes to the hiring system, brought about by changing expectations and economic conditions allowed the yearly bond between the farmers and lads to continue through the 1940s. Although the system was being put under pressure, it was still necessary to the functioning of the farms. In the 1940s the horse was still an integral part of the agricultural life in Holderness, but as tractors improved, the role of the horse diminished.

Just as the hiring fairs were still going strong during the war years, albeit on a smaller scale than when horses were the sole motive power on the farms, the way the farms were organised and the manner of working the horses were still basically the same as a generation earlier. The day-to-day running of the farms was largely in the hands of men who were born around the turn of the century and had come of age working horses. The farming culture still centred on the working horse, so Ron's experience was essentially the same as that of lads from earlier generations. He was still a part of a continuing tradition of farm work in the East Riding, following in the footsteps of countless other horselads.

The East Riding was and is a county of arable farms, and agriculture was by far the biggest source of employment for male workers. Despite this, a career in farming would have seemed unlikely future for Ron Creasey when he was born. For Ron was born in Hull, the son of a docker, and had it not been for the Second World War he might well have followed his father to work on the docks.

CHAPTER 2

From Hull to Holderness

BEFORE the Second World War, the commercial life of Hull and the lives of most of its inhabitants were dominated by the docks. The fish docks, to the west of the city centre, boasted a trawler fleet as big as any in the country, competing with Grimsby for the largest fleet, and when the wind was in the wrong direction the smell permeated the whole town. To the east of the fish docks were the commercial docks, Hull then being the third biggest port in the country, after London and Liverpool.

It was into this world of commercial activity that Ron was born, to Annie and George Creasey. This was in October 1929, when the brief boom years following the First World War were over, overtaken by the poverty and hardship of the depression. While tenant farmers throughout the country were struggling to pay their rent as the prices for their produce tumbled, and new tenants were taking on farms free of rent, the reduction in trade and shipping meant that there were also hard times for those who worked on the docks.

The most important commodities of the Hull trade were coal and the timber coming into the country from the coniferous forests of Scandinavia. Both Ron's father and grandfather worked at the heart of these trades, Ron's father as a coal trimmer, and his grandfather as a foreman deal carrier.

Ron's father worked for the Ellerman Wilson Line, his job as a coal trimmer requiring him to go down into the ships' holds while they were docked, to level the coal before the ships went to sea. Armed with a heart-shaped shovel, an iron plate to stand on, and a tallow candle for light, the coal trimmers worked at any time of day depending on the tides, and emerged black from head to foot with coal dust, often with cuts to their heads from where they had bumped themselves on the roof of the hold. The reason why so much coal passed through Hull was its proximity to the large York-shire coal fields; all the coal used to come into Hull by train.

> When we went to school, you would see all the trains coming in loaded with coal, and all the trains going out were loaded with timber. You'd see it coming in on these ships and it was loadened on the decks. The holds would be full as well, but there was such a list on, you'd think they were going to turn. They were creeping into the Hull docks, stacked right up, ever so high.

Ron's maternal grandfather, George Storr, worked in this part of Hull's trade, and was one of those responsible for carrying and stack-ing the softwood timber after it had been unloaded from the ships or had returned from the sawmills.

> Me grandfather was a foreman deal carrier, the 'deals' was the timber, and it was stacked in stacks, and they were nearly as high as a house. When they were stacking it, they left a plank out so they could put a plank onto it, so they could walk up it, and then there would be another one to go down, and there was relays of men all up these planks. These deal carriers worked for ' railway and you'd see all see all the horses and the timber waggons, and they would be fetching this timber off, and going to the sawmills down Hedon Road. And when me father was on the dole, which was very, very often, we used to walk to the dole office; you never thought of taking a tram, and he would sit me on ' back of these timber waggons when they were empty, and we would go down Hedon Road, and when it turned off we just jumped off and walked up to ' dole office. When you went up to the dole offices, they was queued out of each door right across the road, and they had to go every morning to sign on in them days.

The dock workers never knew from one day to the next if there was going to be any work, so before going to sign on they first had to go to the docks. But they only needed to go as far as the gates, where they could see Ron's uncle, who would gesture to them to indicate if there was any work.

> Me father's brother wore a uniform, brass buttons and proper tippy sailor's cap, a tippy cap as we called it. You'd see him and if he'd just go like that, there was nothing doing, so they'd just turn and walk away. Now if they went to work, they took it in turn; very rare they

was all at work at once. There were so many coal trimmers: there might be four of 'em, and this foreman feller would tell 'em where to go, and then there'd probably be another boat going and there'd be another three go to that. And every year, New Year's, when they got a new almanac, it was turned over, and on the back it was lined out and all the men's names were put down one side, and as they went to work he used to cross it off so everyone got a turn.

It was during these years of the depression that Ron started school, first at Craven Street infants, then the juniors, but by the time he reached the seniors, the Second World War had broken out. After the initial quiet time at the very beginning of the war, Hull became an obvious and important target for the German bombers because of its status as a major port.

Hull was bombed quite a lot. It was supposedly the most bombed city for its size in England; and 1941, when they came for the two day blitz, Hull city centre was knocked right out. That's when half of Hull thought it was time to be out, and we went to the little village of Sproatley, which is about six miles outside of Hull. We went to Lodge Farm, to a Mr Porter; me, Father and Mother. Some people stopped a matter of months, and I think me father and mother stopped there ' best part of a year. So I then went to live with a family, Mrs Beckett, and she had a son and a daughter in the forces and two sons my age and a daughter at home, so I went to live with them.

Moving away from Hull meant that Ron also had to go to the village school in Sproatley, which suddenly had a great many more pupils than previously. Although school discipline at that time was generally strict and corporal punishment was common, this particular schoolmaster was very ready to use his cane at the slightest provocation. As Ron was often at the centre of any mischief-making, he was made to sit at a desk right at the front of the class, within sight and easy reach of the schoolmaster.

When the evacuees went to Sproatley, the school nearly doubled in pupils for a while, and then as the war went on they started to drift back, but I never went back, ever. And the schoolmaster, if it hadn't been for the wartime, he would have been retired, and very often some of the parents complained about this schoolmaster. They said he was the most educated schoolmaster in the whole of the East

Riding, but he didn't seem to have any idea, or he'd lost all interest, and we more or less did as we liked in the school.

And two afternoons a week was gardening. We had big school gardens, and us lads did all these gardens; dug his garden, picked his fruit, dug his taties up, as a lesson, 'cause they thought farm lads was going to be farm people, or gardeners.

And he was very handy at caning us was this man. My last year there, I think I got six raps o' ' cane nearly every day I was there, for saying something or doing something. He just used to get this cane, and didn't he use to belt you with it! He used to stand on the platform where his desk was to get more leverage at our hands, but as he swiped we used to pull our hand away and he used to fall off the platform! And another thing was we held our hand out over the piano and then pull our hand away and it rattled the piano, and we got up to all these stunts!

Well, this particular Friday afternoon, we took this cane, it was a bamboo walking stick cane. He allus kept it in the corner in a waste-paper basket, and we knew what was going to happen, so we got the cane and hid it, and we was swilling this yard. He was allus wandering about, and he came and we said, 'The tap won't work, sir.'

'The tap won't work?'

'No.' And we was looking down the hose pipe, and he gets it off us, and he was looking down the hose pipe, and then we give them the signal to switch the tap on! Well, hell, he was infuriated! So he runs in the school, and goes for ' cane. No cane there, cause we'd hidden it; he rushes back into his own house and he comes back with this brand-new cane, and I was the first one he met, and by, he hit me with it! He didn't hit me with the cane, he hit me with the handle and it stuck in the palm of me hand, and then he gathered us all up, and didn't he welt us all. And if there was any smaller ones, we used to shove them at the front, because by the time he'd got to us, he'd tired out!

And he had a dog, and one desk we was sat at, there were some cupboards at the back of us, so we 'ticed the dog, and got it in the cupboard and shut it in. And after a while, he'd suddenly gaze about and couldn't find his dog. 'Anyone seen my dog?' No-one had seen the dog; and after a palaver about the dog, it started to get restless in the cupboard, so we slid the door open and the dog shot out! We was a candidate for the cane then!

It was not just at school that Ron found himself involved in, and sometimes responsible for, mischief. Outside of the classroom, window tapping was a common source of amusement for the lads as well as a method of annoying the neighbours.

And then we used to go window tapping at night. You had a long length of cotton, a pin and a button, and you stuck the pin in somebody's window and the button was on a piece of cotton probably six inches long, and when you got at ' back of a hedge and pulled the long piece of cotton, the button would tap on the window. 'Course they would open the front door and look out and see nothing, and go back in ' house, then we'd tap again. Well, most of the people in the village knew what was window tapping, but some of the houses would chase you. Well, if they didn't chase you, we didn't bother them no more; we went to the houses that would aggravate people, where we could get a run.

And this man, he was a Londoner but he'd lived in the village about twenty years, so we knew Arthur was good for a run. We goes to this cottage, and we started to tap at this window, and there were little square windows and this lad's fist went through the window! Well, be that, we were off down the road and this London feller after us. And I wore clogs: it wasn't a clog area, but clogs weren't on coupons, so clogs we wore; wooden soled with a horse shoe type of iron on the soles so when you were running down the road it was like a pony running. So down the road we goes, Arthur fast after us. And when we got to where I was living, we all rushed in the house. She said, 'Get upstairs.' Took me clogs off, and Arthur knocks at the door, 'Is that lad in, Missus?'

She said, 'No, he isn't, Arthur. Why?'

'Well, they broke my window.'

'Well,' she said, 'his clogs are here,' and by this time we were all looking out ' the bedroom window down at Arthur.

'Oh, I'll go and find one of the other bloody lads,' and off he goes . . . and from that day on they called me Cloggy.

In the 1940s village life was still intimately connected with farm life, and even during school time the older children would sometimes go to work on the local farms to help with labour-intensive jobs in the potato fields, and for singling mangels, known locally as *wuzzels*. These big roots, which were mainly grown to feed cattle, have a multiple seed, so three or more plants emerge from one seed, and it was the lads' job to remove all but one of these seedlings, to leave one big strong plant.

These farmers used to come to the school and say they wanted so many lads for tatie picking or singling wuzzels, and we'd all troop out into the cloakroom and this feller'd be there telling us what he wanted and when he wanted us. Then we'd all troop back into school and when the day come, off we would go.

When we used to go to farms tatie planting, they used to have so many rows made when you got there. We used to start about eight, 'til five, and we used to have our dinners packed up, but very often by ' time we'd walked to the farm, which could have been two and a half miles, we'd eaten our dinners before we got there in a morning, and so when it come to be dinnertime we'd nothing.

Five school lads used to plant four acres a day, and a pair of horses could make four acres and more a day. There was a cart in front of you loadened with taties and a man sat in it, and as you emptied your bucket, you handed the empty one up to him and he handed you a full one, so that was his job to keep all the five lads in wi' buckets and keep 'em going. That's how you got the four acres, whereas some places you went, they'd got 'em laid out in *bags*. But you'd to stop planting to fill your bucket; well you lost a lot of time.

When we first went to John Caley's tatie planting, he used to plant eight acres at Carr Farm and eight acres at the other farm. So we was there two days at each farm. Then you'd to go for your money, and Foreman used to say, 'You go and see him for your money!' So we'd go, and he used to be in britches and leggings, and he used to pull out these handfuls of notes; we'd never seen so much money in our lives. Out of this pocket he'd pull handfuls of silver.

'How much do I owe you lads?' And you'd say, 'Well, we've been here two days and that's four shillings a day, and we've been two days at Pasture House.'

'Oh, so it's sixteen shillings.' And this is when all these handfuls of notes come out, and he generally give us the pound, give us the extra, allus. He was a good feller.

In addition to this seasonal farm work, from the time Ron was first evacuated from Hull, he also started to work after school on the farm where he lived.

When I was twelve at Porter's I got a job every night. I fed ' pigs and got the cows in. There was two hundred bacon pigs to feed and seven sows, and then there was twenty cows and you used to shout o' them and get them in, chain 'em all up and give 'em their bit of meal, and that carried on 'til I was fourteen.

And every day we'd to go and get a lot of kale and take it and spread it on the grass. Then at night we turned the cows out and they went down to the pond, because the only water there was in any of the fields was ponds; and they went down and got a drink and then they ate the kale and then they came back up. By this time you'd to have the cows mucked out and all bedded up and the racks full. In a morning the racks were full of oat straw, but at night it was full of clover hay. So when they come back up they were all waiting at the gate, you would be opening the gate and letting 'em come back in, and the farmer hisself milked them by hand. He was milking 'em in a morning when we got there, and he didn't start milking them after tea *while* about six, and he used to say he could milk them as fast as any milking machine but he couldn't milk three at once. And then eventually we got Alfa-Laval milking machines, and then they got another woman to look after the cows.

Besides working on Porter's Farm, Ron also saw the work on other farms, one of which was Pasture House, where he had gone to plant potatoes. This was home to his friend Harry Buck, whose father Charlie had been foreman there since the late 1920s. Like many other school lads growing up on and around farms, Ron had the opportunity to try doing different jobs when time was not pressing, and for those like Charlie Buck who took a great interest in farming, giving lads a chance to have a go was part of their early training.

On a Saturday morning, he would say, 'If you want to come down, call and get me a bottle of tea,' and you used to go to Pasture House and she'd give you a bottle of tea and a sandwich, and then you'd go down to where they were ploughing. There'd be Waggoner and Thod Lad and Bullocky Lad; there'd be four of 'em ploughing, and then he'd say, 'Do you want a go then?' And he'd give you hold of his plough and he'd stand on ' *headland* . . . oh, he'd got a good pair of horses.

Although the horses knew very well what to do and the plough was already adjusted, this sort of early experience meant that many school boys were familiar with many aspects of farm work, even though they were not particularly skilled. Before ever getting a job, they had already learnt that working on a farm involved both hard work and skill, but also that people took pride in doing a job well, so if they concentrated on their work and strove to increase their skills, they would be respected in the community.

For many routine jobs farm children were expected to work largely unsupervised, so when feeding the pigs at Porter's, Ron was left on his own, only calling on someone more experienced when something was not as it should be. Naturally, the more demanding work was done by more experienced workers, whether ploughing with the tractor or drilling the corn with horses, but Ron also worked with the horses, especially when there was someone else working with him.

> All the carting, the drilling, and the rolling was done by horses, and so by ' time you was twelve, you drove horses. By ' time I was twelve, I had one 'oss in front of the other *leading* wuzzels off, and the odd time I would take one and loaden a load of sheaves. But by ' time I was thirteen I'd a *rulley* all the harvest; I had one permanent, all the holidays.

These early experiences were formative for Ron, sparking an interest and passion for farming, and especially for the horses, which he retained throughout his life. Although he was brought up in the town, once he was on farms and around horses, he was in his element.

CHAPTER 3

Lodge Farm

IN the 1940s the school leaving age was fourteen, but having worked through harvest on Mr Porter's farm when he was thirteen, Ron simply carried on working instead of returning to school. This was in 1943, when most young men where away at the war, so except for Sid Porter and his wife, all the farm staff were under eighteen. None of the workers at Lodge Farm were hired at the hiring fair, but instead lived at home in the village, except for Ron who lodged with the sister of one of the lads. Although the lads did work the horses, they were also involved with all the other work as well.

Lodge Farm was approximately two hundred and fifty acres, situated at the edge of the village, and only a field away from one of the lodge gates to Burton Constable Hall. Sid Porter was known in the district as a good farmer and was a hard worker, starting his day by milking the Dairy Shorthorn cows before the others arrived in the morning, except on rainy days when everyone had to sit and milk one. Apart from market day on Monday, and the occasional visit to Beverley market on Wednesday, he was always at work on the farm. Not only did he work hard himself, he also expected everyone else to do so, and even to the young lads he would say, 'If you can't talk and work, don't talk!'

The farm included large areas of grass in the park land around Burton Constable Hall, though some had been ploughed up as part of the war effort to produce more food. But much of the farm, as was typical in Holderness, was under arable crops.

There was a field of mixed corn, which was barley, oats and grey peas, and the grey peas grew up the barley. That was for feed for the cows. But there was also barley. I know there was sixty acres of wheat because we always *sew* the wheat in the autumn, and you'd

Ted Simpson with a pair of horses at an unknown location in Holderness, pre-1914.

Ted Simpson's brother with a horse in cart harness, pre-1914.

Ron, aged three, sitting on his uncle's rulley.

'Luance time when muck leading at Porter's in 1945. From left to right: Ron (aged 15), Harold Robinson, George Riley, and the farmer's son David Porter.

Ron with Jet and Violet at Carr Farm.

Ron riding Jet, with Violet tied to Jet's hame ring. Note the cord neatly coiled from the leather loop through the hame ring.

Tim, the Third lad, at Carr Farm.

Third Lad in the stackyard at Carr Farm.

Ron leading muck in a waggon at Pasture House with Violet and Belle.

12 October 1948: Flinton horses at Pasture House stackyard. Royal is the nearside wheeler, with Belle on the other side of the pole. Cobby is the nearside leader, with Turpin next to him.

Ron at Pasture House. The check rein going to Cobby is clearly visible.

Ron and
Maureen Briggs
riding Paddy.

Ron at the
Cottingham Show,
August bank
holiday 1949, with
horses belonging to
Laurie Caley.

Ron with one of the Scottish horses shortly after the sale.

Four horses in the stackyard at West Newton about 1951.

bit of a struggle getting it in. And we always had twelve acres of peas; some on 'em was blue peas, but we also grew grey peas so we could grind it for cattle feed. And at times they had to be mown by scythe, 'cause they were that flat you couldn't cut 'em with a *grass reaper*, and so everybody had a scythe. There was the farmer hisself, his son and two more lads and they were all below eighteen, and then there was me at fourteen. Then we got a land army girl, she was sister to these other two lads. And at harvest they got extra help, people who had jobs, shift work, they'd come.

Then we had three or four acres of kale. Nobody liked getting kale, ooh, it was a dreadful job; you got wet through. You couldn't get wellingtons, because you had to have a certain job for to have rubber boots. So we had our boots and leggin's and there was no waterproof clothes, and when it was wet you used to be wet through. And when it was frosty, when you hit the stalks of kale with your bill hook they used to splinter, fall into little bits. We always loadened it with our hands because that was the quickest, and you put it on all straight, with all the leaves looking outwards and the stalks inside. I used to get help some days, and some days you didn't, but by the next year, then I was fifteen, you were getting stronger and this kale wasn't such a bad a job as it was when I was fourteen.

And the last year we had this kale, every day when I went for it when it was nice, I used to think, 'Well, that's a load,' and then I would cut some more and put it on. Very often on these farms, you weren't supervised that much, they give you a job and you had a certain time to do it in. And when it came to Christmas, he said, 'Get plenty of kale, Ronnie.'

I said, 'There isn't none left,' and I'd cut into it and cut into it 'til there was none left at Christmas. But the previous years, it was there 'til the spring. It went to seed 'til we'd to plough it in.

And of course we had our two acres or three acres of mangels, or wuzzels. The mangels were all in a *pie*, and at that place we always used to pie the mangels in ' stackyard. It was traditional, three cartloads wide and taken up to roughly one mangel wide at the top and then we used to straw them down. It was all loose straw, put on in a particular way. It wasn't left to anybody to put it on, it was put on by somebody who was a bit more responsible.

On many farms the mangels were clamped, or pie-ed, in the field where they grew, and a layer of soil put on top of the straw to keep the frost out over winter. But building the mangel pie in the stoned stackyard meant that there was no soil available, so the whole of the pie was covered with mixture of hay, thistles and brambles that had been cut on the *dyke* banks.

During the winter these dyke *reapings* and straw were taken off one end of the mangel pie to get at the mangels so they could be fed to the stock. The cows got their mangels whole, but those for the fattening bullocks were first put through a root cutter, which was turned by hand. The resulting cut roots, like big chips, were then mixed with other ingredients before being fed to the bullocks in big wooden troughs known as *tumbrils*.

> We pie-ed these wuzzels quite near the barn and then we used to take the wuzzel cutter, or turnip cutters as they're called, to the heap, and we used to cut there and then carry 'em in *scuttles*. And when we mixed all the rations up for the bullocks, you put a layer of *chaff* and then a layer of mangels, and then a layer of *sharps*, then threw bucketfuls of treacle water on, and then you start again with your chaff and your mangels and we had a heap maybe three foot deep. And then it used to heat up, it used to heat up very quickly, and if your hands were cold, you could shove your hands in the heap and get your hands warm.
>
> Then we carried this out in little *sheets*, and you'd to put so many sheetfuls in every tumbril for your bullocks. There was always two of you feeding these bullocks, one filling the sheets and helping you to lift 'em on. Of course the treacle used to run out and run down your jacket a bit, but nobody bothered too much in them days.

In the war years, agricultural working hours were from seven in the morning until five at night, with one break in the middle of the day. In Holderness no-one stopped in the morning or afternoon for a drink or a bite to eat, except when doing specific jobs, at hay time, harvest, when mucking out the fold yards, and on threshing days. On these occasions the allowance, or *'luance*, was provided by the employer. Besides stopping work for 'luance, on threshing days everyone also started an hour earlier in the morning, in order to feed the livestock before the threshing started.

> We used to get paid to go at six o'clock on threshing mornings, 'cause you was paying for the threshing machine by the day, so we got all bullocks fed and ' pigs fed before we started. At Porter's we

used to be threshing at seven and we'd be threshing 'til half past five. He'd either be on ' corn stack or on the straw stack, and when he was on the straw stack, when it was coming up to 'luance time at morning, you'd see him making this massive forkful of straw. Mrs would fetch 'luance into ' yard and then he would stick his *jack straw* into this forkful, and ' straw stack were just high enough for him to come over and land on his feet, and he'd take this forkful and bed about ten cows up.

Except on a few of the very biggest farms, most farmers did not have their own threshing machines, but instead relied on a threshing contractor who served the farms in the district. Besides the contractor, and maybe one or two men who worked for him, the rest of the threshing gang was made up of the farm staff or casual workers who followed the threshing machine round the different farms.

The men who was with the threshing machine came in the night before and set the machine up, and it was steam at this time, so he slacked his fire down and next morning they came and opened their firebox; then they came in for their breakfasts. They went in for their breakfast, dinner and tea there, if they were stopping, 'cause sometimes we had seven days threshing at a time when we got going, but not always. When it was early, after harvest, everybody wanted to thresh at once and they [the farmers] was threshing for money, that was the first draw they'd had. But they also wanted straw so they could cover their stack tops down. So they'd give you one day and then move out that night, and it was very often dark, and they'd go to the next farm and they'd set up there, so they was ready to thresh next morning.

Billy Gibson was the threshing machine man and he kept his threshing machines at a village called Marton. He did a radius of round Marton, and they was threshing nearly every day when it wasn't too windy or it wasn't wet, all through the winter. And I remember Sid Porter saying to Billy Gibson, he used to pay him in a cheque did Sid, and I happened to be stood by ' traction engine and he says, 'You'll knock that off, Billy, for finishing early, won't you?'

We finished early which was very rare, and Billy said, 'Yes, I'll knock off for what you give me extra for that one when we were threshing late!'

After the threshing machine had gone, there was a big stack of straw in the stackyard for every corn stack that had been threshed.

On many farms, the straw was taken into the fold yards to bed up the bullocks with horses and a waggon, but at Lodge Farm the straw was usually carried with a big fork, known as a jack straw.

> There was several jack straws on that place, you could put it through a child, put it under his arms and carry him in this fork. At that place we used to carry straw into ' yards in a morning: we didn't always have horses. The only time we did have horses there was at ' weekend. We'd carry in, and you could carry some straw, one man making forkfuls, so when you'd come back, you'd stick your fork in and he'd help you to rear 'em up. Once you got it on your fork, it was straight down your leg, and you couldn't see where you were going; you couldn't see forward, but you could see where the straw had fallen out from the last time. You followed that track, and them big straw doors were quite wide and they had a chain on, and as you went through you pulled this chain at ' back of you and when you come back, maybe ' other feller'd be coming in, so you'd undo it for him and you'd go off for your next lot. There used to be about three of us bedding these bullocks up. We carried straw into the yards for bedding and then we took one of these big forkfuls and put 'em in every tumbril for 'em to pull at; one day it would be barley straw then another day it might be pea straw. So you had three or four straw stacks going, and they had a variety of pea straw or barley straw and occasionally oat straw, so the bullocks always had a variation.

During the war years there were many farmers who relied solely on horse power, but many others, like Sid Porter, used both horses and tractors. The tractors first took over the jobs requiring the most power, such as ploughing, and also the cutting of grass and corn, which were always hard work for the horses, particularly when the weather was hot. Lodge Farm had one tractor, a Standard Fordson, and the rest of the work was done by five work horses. One of the limitations of the tractor was the iron wheels with lugs, which pushed into the soil to provide traction but prevented the tractor from being used on the road. So all the road work, as well as the haulage on the farm was done with horses. Although in other parts of the country two-wheeled carts were very common, in the East Riding the most common farm vehicle was the Yorkshire pole waggon, which was used most frequently with two horses, one on each side of the pole. At Lodge Farm, however, there were no waggons as Mr Porter preferred to use rubber-tyred rulleys, four-wheeled vehicles without sides, which had shafts for a single horse.

As with most tasks the young work force was largely unsupervised when working the horses, but they were still expected to get the job done. While usually uneventful, things could go wrong. One such instance occurred when Ron had to take a load of muck through some park land, which had been ploughed into *lands*, high ridges with furrows in between, to allow for better drainage.

> There was three rulleys and a cart, and rulleys were a lot better than carts, 'cause with a cart you're allus stood on your muck, you're allus paddling it down, whereas on a rulley you could stand side by it. And I'd been told off for getting behind, I'd been smoking, and every time I opened me mouth to say, 'Go on,' the bloody cig flew out ' me mouth. I had to get out the cart and pick the cig up and get back up, and I'd got behind. I'd had a good telling off about this, so I was going to catch up. It was in park land, the fields were landed up, they must have been landed up first and then the trees planted, and these trees were all somewhere on the land tops, and some weren't.
>
> And I thought, 'I'll cut through this planting.' It was just a stand of trees, no fence round it or anything; save a bit of time. I was sat on the cart, and suddenly this horse decided it would side-step and there was a sapling and it went straight between the horse and the shafts, and with us going down this land, the sapling went as far as the shoulder chain would let it go. But with a load of muck, there was the cart at this angle, the horse couldn't back it. I couldn't turn it side-ways because it was *fast* with the sapling. All the others were at ' far end, and I'd to wave and shout. Then we'd to take the *etch* off, tip the cart up, get the horse out, and then lift the shafts up the sapling and then twist the cart and put the shafts down; put the horse back in. Then we'd the load of muck to put back in the cart. Then didn't I get me name for nothing!

Being told off for not concentrating on the job was a fairly common occurrence especially with young lads working together. Another occasion when Ron and one of the others risked the wrath of Mr Porter was when they had been sent to spread slag, a by-product of steel-making used as fertiliser. All the slag came in hessian sacks, each one of which had to be emptied into the fertiliser drill, then shaken out, folded and bagged up.

We were all lads, I'd be fourteen, and there was two of us there, one used to drill it, the other one was opening the bags, and it poured down with rain. We were sat under this rulley, and we thought, 'Hell, it isn't going to stop.' So we *loused out*, 'cause we'd one 'oss in the drill and one 'oss in the rulley so we could keep moving down the field, and we were riding merrily home and we got nearly to ' farm and he says, 'What are you doing?' And it had stopped raining, but we hadn't noticed.

'Well,' we said, 'it's raining.'

'Well,' he said, 'it isn't,' and it wasn't, so we'd to turn around and go back. But I suppose he understood that we were all lads, whereas if we'd been men, that would have never happened.

The use of basic slag and other bought-in manures increased greatly during the war, in part because during the depression years much of the land throughout the country had been allowed to deteriorate. Thousands of acres of arable land had been allowed to revert to grass, but at least it was storing fertility in the sward, which was cashed in when ploughed up to grow corn during the war. The land still under arable crops, however, was either low in nutrients or had become acidic through continuous cropping without adding lime.

During the war a lot of the fields were very deficient. They'd been let go through the bad years, and farming was picking up again, and this particular field was deficient of lime and it wanted three ton to the acre. In those days it used to come in bags from Earl's cement and they were eight-stone bags tied with a wire clip. Earl's cement would bring it and if you wanted three ton to the acre, they put a ton on the first headland, drove across the field and put a ton on the middle and a ton on the other headland. And when you come to drill it with ' *artificial* drill, you filled the drill up and put so many bags on the lid, went halfway across and then stopped and put the bags off the lid into the drill, and then you reached your middle heap. And then the same thing again 'til you reached the end heap. That's how you got these big amounts on. You were forever filling, and the little lad had to shake the bags out, turn 'em inside out and then bag 'em all up again, 'cause all the bags went back. They were all hessian bags, rather a big-weave hessian bag, and everybody was white. Everything was white, the horse, the men, and ' course it wasn't long after that that they got spreaders so the bags were done away with. So it didn't really matter how many ton they put on, it was no hardship to anybody then.

While the lime was delivered to the farm, other things were collected from the railway station with a horse and rulley.

> We used to fetch all sorts off ' station in those days. We used to fetch salt; that used to come loose and we used to bag so many bags up and put them round the edges of the rulleys, and then we used to have to shovel it all on, loaden 'em right up.
>
> And also the seed taties used to come from Scotland, and we used to go to the railway station, usually Hedon when the taties used to come. Then we used to fetch 'em off, and there was always oat straw lining the railway waggons, beautiful oat straw, we'd never seen straw so long and so green in our lives. Then we used to fetch basic slag, it was all in little bags, but by, they were heavy. There was twelve stone in a bag tied with a little wire tie-er.
>
> And when we were drilling slag there, at one particular point there was a lad in the field with the drill, two more people at the farm and they used to loaden the rulleys and they'd put a ton and half on [per acre], and I had a horse and I used to take this ton and half to the lad drilling, louse the horse out, leave the loadened rulley in the field, put it in the empty rulley, and take it back. Then when I got back to the farm they'd already loadened a rulley with a horse in. I backed this one into ' shed then took the other one and by time I'd got back he'd got that ton and half drilled. So all I was doing was just driving up and down.

Applying fertiliser in powder form was often a difficult job in Holderness, with its exposure to the fierce winds off the North Sea, so when the wind got too high, the work would stop and a sheet was put over the bags of slag to keep it dry.

> If ' wind was wrong and it was blowing it into ' next field, you'd stop and have to come home, but if the wind was blowing it in front of you and it was staying in the same field you carried on. But if you had to sheet down, there'd be two of you holding the sheet and it'd be stood like a sail up in the air, and we couldn't get 'em down.

The wind was also a problem at harvest. Not only did it make stacking difficult, but the wind often flattened the peas and the corn before they were ready to cut. The peas were usually cut with a mower, but when the crop was *laid*, often it could only be cut in one direction so that the cutter bar could get underneath to the bottom of the pea vines, rather than riding up over the top. This meant that instead of cutting round and round the field which is the quickest way to do it, the mower, or *grass reaper*, would have to go back to the other end of the field without cutting anything, *fetching* it, as it was called. The corn was usually cut with a binder, but if necessary, it was fetched as well.

> You did the same with the corn, running down just one side and then running back empty, or you might be running down two sides. In the '40s, when ' weather was bad, we was cutting corn with the grass reapers, and that had to be moved to let the reaper come round again, and also they had scythes out and they were mowing. I've seen 'em mow twelve acre of peas with scythes because they was so flat to the ground they couldn't get the grass reaper.

Even if the corn was standing, the first part of harvest was still hand work, cutting round the outside of the field with a scythe to give the horses somewhere to walk when they were making the first round with the binder, and making space inside the gate to allow the binder to be moved from its transport position which let it pass through the gateway, into its wider working position. Once the binder had cut the corn, the sheaves were all stood up in *stooks*, to get the heads containing the grain off the ground so that the corn would dry.

> We used to stook with our shirt sleeves down. Before this, when there were thistles we used to go hoeing, but very often we didn't get round all the fields, and some of them had a lot of thistles in and if you were seen picking 'em out the field they'd say, 'Don't be doing that. Leave 'em all in, or you'll have nothing to do at ' weekend!' And then when we came to *lead*, we were loadening by hand and then same thing again, some of the fields were that bad you were loadening with forks, and I remember once stooking with forks. There was that many thistles in it we couldn't handle it.

After stooking, the corn was taken to the stacks, with a number of vehicles travelling between the field and the stackyard. Carrying the sheaves of corn from the field, known as leading, was usually done with a pair of horses pulling a pole waggon; but with good reason, Sid Porter preferred using rulleys.

They didn't run pairs there, they only run single horses, and all their rulleys were on rubber tyres, because he said a horse with a rubber-tyred rulley could pull a lot more on the land than one with iron wheels. When it was wet, we could come off with loads, where we could see over the hedge where the farms with pairs of horses were having a struggle with their waggons. But when it got real wet of course we had *fost 'osses* on; two horses, but one in front of the other.

To organise the work so that everyone was kept busy, both on the stack and in the fields, a succession of three or four rulleys followed each other between field and stackyard, each of the lads staying with their own rulley throughout the day. In the field there was a man who forked the sheaves up onto the rulley, and a little boy who would fork the first few layers, or *courses*, when an empty rulley came into the field. To get the work done more quickly sometimes the small boy might swap places with the bigger lad on the rulley until the man had finished forking the previous load. With a crop of wheat, each waggon load held about forty stooks, and twenty-four of these loads would go into a stack, which was the crop from at least ten acres, each stack being a day's threshing. At the stackyard, the lad with the rulley usually got up on his load and forked the sheaves onto the stack, *teaming*, as it's known in Yorkshire. There were usually three people on the stack, the picker who forked the sheaves away from where they landed on the stack; the stacker, who stacked the sheaves in their final position; and the middle man, who forked the sheaves to the stacker, turning them so they landed handily and the right way round for the stacker. When Ron was still a schoolboy, however, instead of him forking the whole load, he would swap places with someone on the stack so he did not have to fork the sheaves uphill when the stack became higher than the load he was standing on.

When the stacks got high, a little before the stacker started topping up, drawing in the top to make it roof-shaped to shed the rain, he would stop stacking where the picker was standing, to leave a small space called a pick hole, large enough to stand in and fork the sheaves upwards. To prevent the lad slipping out, as all the sheaves in a stack are angled outwards, a pot picker, which was usually used to find the position of land drains, was pushed down into the stack for the picker to stand on.

You were stood in this hole with your toes on this pot picker, and they stacked up at ' back of you, and when I was at Porters, as soon as he got near you he'd tap you on the head with ' fork and say, 'Change hands,' and you'd to change hands and go that way, so when he got to ' other side, he was left-handed and you'd be going right-handed. You couldn't see what you were doing and it wasn't only straight up. They started to top 'em up as well, so they were going away from you. And then you'd start to slide out and ' course, when the rulley pulled away, there was nothing there, you were stood looking down, waiting on the next one coming, and you'd feel your feet slipping. You'd say, 'I'm slipping, Mr Porter!'

'You'll be alright. You'll be alright!' He'd say.

Harvest was one of the few times in the year when there was any overtime. Before the war when money was tight, the farmers did not want to pay overtime, even if there was work to be done. But at harvest the need to bring in the corn, which was the farmer's main source of income, meant that the work continued until dark. The war also brought additional work because of the increased arable acreage to boost food production, and despite being in a reserved occupation, many younger farm workers had joined the forces, further exacerbating the labour problem. One measure which helped the farmers get in the harvest was the government's introduction of double summer time, the clocks being put forward two hours ahead of Greenwich Mean Time. Despite having this extra hour of light in the evening, Mr Porter still wanted to get more loads out of the field.

We used to loaden all the *draughts* up at night. There was never any left empty in case it rained, and then there was something to do next morning while it got a bit dryer.

At harvest we worked 'til dark, and very often at harvest at that place, he would say, 'Go for a moonlighter!' The moon was out and off we'd go, and when you got there the moon would go in; you couldn't see where you were going, and when you were coming out [of the field], one of you stood at one gatepost and one at the other, and told the other to come, and you got all the three vehicles out through the gate. Then we'd go home and you could see the sparks flying off the horses' shoes. And we was harvesting 'til eleven o'clock at night.

Of course we was now earning more in overtime in a week than we was getting wages. And we used to get 'luance twice. They used to bring 'luance out at four o'clock, and then bring ' luance out again at about half past seven.

Another wartime change to the farming in Holderness was the introduction of potatoes. Before the war no potatoes were grown in the area, the clay soils making it hard to create a deep tilth for planting potatoes, and making harvesting very difficult once the land became wet. Nonetheless all the farmers were compelled to grow a certain acreage of potatoes, and although the conditions weren't ideal, once they found out how to grow them, they continued to do so after the war.

At this time, the war had come to an end and Germany was starving. So everyone who grew taties had to leave so many yards of tatie pie 'til the government told them they wanted them [to send them to Germany]. And when they wanted them you had to *riddle* them, and the train used to come up into Ellerby station, which is only about three miles away, and all the district had to take their taties and put 'em on this particular train within a stated time.

And these tatie pies, the leaves were grown right through the soil, and we used to take the soil off, and then they were all grown through the straw so you'd to pull the straw off, and knock all the taties off, and then we used to riddle them once with no bags, and they fell out at the other end making a heap, and then we used to turn around and riddle 'em back again. We used to have about ten ton to put on and they were all bagged up in eight stone bags with MF [Ministry of Food] on them, and they all had to be double sewn. Then we'd two pair of horses, we'd put four ton on each load and I used to go with one; we used to put 'em on the railway station.

Working for a good farmer like Sid Porter was an excellent start for Ron, but after three years' experience of general farm work, he was ready for a change and the chance to earn more money working full-time with horses. So within a fortnight of leaving Lodge Farm, after being hired at the hiring fair in Hull, Ron was on his way to work at Carr Farm, Flinton, which was run by Sid Porter's brother-in-law, John Caley.

CHAPTER 4

All in a Day's Work

RON's new boss, John Caley, was part of a big farming family, and most of his brothers also farmed in the area. Their father, Harry Caley, had started farming a couple of miles away, and as his sons became old enough they took over farms as they became available. John Caley lived at Manor Farm, which he ran as one unit with the adjacent Carr Farm, but he also managed Pasture House Farm, where Ron had planted potatoes and had his first go at ploughing as a schoolboy.

Although John Caley was actively involved in the running of the farms, at Carr Farm the man directly in charge of the horselads was the foreman, George Gibson, a man in his late 40s, who commanded respect and expected everything to be done in the proper manner. Carr Farm covered over three hundred acres, added to the one hundred and twenty acres of Manor Farm, but in those times people did not refer to a farm in terms of number of acres, but as a six-horse place, an eight-horse place, or a ten-horse place, to give an indication of how much work there was, and therefore the approximate size of the farm. As was typical in Holderness, the soil was mostly heavy clay, which was always difficult land to work, especially in a wet time when the tractors would get bogged down and the horses came into their own. Although there had been tractors at Carr Farm since 1936, in 1946 the horses were still central to the operations, though compared with the years before the tractor's arrival they were now fewer in number.

In Ron's time the ten work horses were all stabled at Carr Farm, although before the tractors' arrival another six were also kept at Manor Farm. The stable at Carr Farm had both double and single stalls, so that there were five horses on each side of the door which opened out onto the road, with another door in the opposite wall leading into the fold yard. The stable was the centre of the farm; it was where everyone congregated at the start of the day, and the centre of the horse lads' activity, not only when they were working, but also after they had finished their day's work.

When the lads arrived at their new farms on the Sunday evening at the end of Martinmas week, traditionally around tea time, they were shown their horses and started work before going in for tea. But when Ron was hired many of the lads, including Ron, did not arrive until later, when they were shown their horses and then went straight to bed, leaving them no time to sort themselves out before their early start the following morning. Most of the lads had never been to the farm before, nor had they met the people they would be working with. It was all new, so Ron's first experience would be typical.

> When we got there, maybe nine or ten o'clock at night, you'd never been before, so you knocked at the door.
>
> 'Hello, come in.' You'd met Foreman maybe when he hired you, but not always, 'cause sometimes Boss had hired you. But you went there and first thing he did, 'Well, I'll just show you your horses.'
>
> So he lit a lamp, took you into stable and there was ten horses in a line and he'd say, 'There's Wag's 'osses, and there's Thod Lad's 'osses, there look, and Fowth Lad's 'osses, and when I tell you to fetch a 'oss out, I want that one, not that one or that one.' And he'd turn around and hold his lamp up, and he says, 'There's the harness,' or the *gearing*, as they called it, and, 'Here's the forks, look!' Your fork was the fork you was going to loaden straw with. 'Now that's your fork, you take it with you wherever you go and you put it back in here. And these are the muck forks and they belong in the stable.'
>
> So then you went in the house, and he'd take you upstairs. There was only us three. Me and Third Lad were going to sleep together, and Wag'd sleep on his own. The bedrooms just had double beds in and a chair against each bed. There was no other furniture, and there was nothing on the floor and no curtains. So they showed you the beds. They said, 'You don't take more than one lamp up here. You put your shoes under there. You don't go upstairs in your boots or your shoes. Your working clothes goes in there,' like a wash house, and he'd give you the rules and regulations and, 'All the swearing stops at that door,' and then that was it.

The accommodation Ron describes was typical of the farms in the East Riding. The lads were not there as part of the foreman's family.

Although they ate their food in the kitchen, their bedroom was separate from the rest of the house, often reached by its own flight of stairs from the kitchen, or up a ladder and through a trapdoor. The lads were expected to sleep in double beds, according to their place in the hierarchy, the waggoner with the third lad, fourth lad with the fifth lad and so on until the least lad, unless there was a bullocky lad, in which case he shared with the waggoner, or if there was an odd number of lads, the waggoner had a bed to himself.

Although there was usually plenty of food, there were no home comforts, and any little items a lad wanted to bring with him from home were kept in his wooden box in the bedroom. These basic conditions were the same on all the farms, and even if the bedrooms were spartan, the lads were not going to spend any time there except when they went to bed. The rest of the day was organised around the needs of the horses, so the lads were doing their horses before the labourers arrived in the morning and after they had gone home to their cottages at night, so that both horses and men would be at work for the whole working day. The lads were always busy, there was always something to do, and there was a right way to do it. If a new lad did not know that on the Sunday night, he would soon find out early the following morning.

> Quarter past five, Monday morning. Foreman give you a shout, 'Let's have yer!' He lit his lamp, did Wag, and if you wasn't sharp enough he'd gone down the stairs, and you'd just hear Foreman telling Wagg'ner what to *gear* the horses for. He might say, 'Gear 'em all for waggons,' or to plough, or whatever. You all slept in your shirts and you didn't wear no underpants, you just had your trousers and your socks, and when he took the lamp, you was left in the dark. Well, the first morning a lot of them had bought their new clothes, and you was left up there in the dark finding your clothes.
>
> When you got down there, Foreman, this foreman in particular, would set a great big fire, you'd think it was ten o'clock at night, and he used to have his feet up on the mantle piece. You got your boots, put your boots on, never laced 'em, went into ' wash house, put your cap and jacket on, and into ' stable.

By the time Ron had got to the stable, the waggoner and third lad had both taken a paraffin lamp from where they were kept in the wash house, lit them, and pulled them up on pulleys high up in the stable. The first priority was to feed the horses, to give them the maximum time to digest their food before starting work. This was not a job to be entrusted to an inexperienced fourth lad, so was divided between the waggoner and the third lad.

Wag fed five horses, Third Lad fed five horses and they mucked three each out, cleaned three and harnessed three. Now I didn't feed any, but I had to muck four out, clean four and harness four. And when they fed them, as soon as they got 'em all fed, they laced their boots up, and I laced mine up as soon as I got in there.

As soon as they fed two, they let two out for a drink and fed their other two. It was all oats and wheat chaff, you just sieved it a little bit and got the dust out, and then you put your whole oats on top of your chaff and fed each horse. Now once you'd got a feed into each horse, Waggoner would let two go out for a drink. They all had *helters* with *clogs* on. You threw your clog into the crib, fastened your *helter shank* round the horse's neck, and he'd let two go, and third lad'd let two go and fourth lad'd let two go, and they were walking through these doors into the bullock yards. That's where the water troughs were.

While they were out, you got your fork and you threw your clean straw forward under the crib and the mucky straw and the muck straight out ' the door. You had to make sure it was far enough from the door, but it was making bedding for the bullocks and it didn't matter how much straw you threw out, you weren't too bothered. Then them two would come back, and then the next two would go out. So they turned three out, and I turned four out and mucked out, and then when they come back they were all fastened up. But you didn't fasten them up with the clog this time, you just fastened them up to the ring, because the clog was only for when they were sleeping; when the horse stands up the clog goes down, and when he lays down the clog goes up, so he doesn't get his legs over.

So they were all eating now, and then you would start to clean them and you had a curry comb, and depending what you'd been doing, you used the curry comb on the horses. You didn't bother too much with their manes and below their knees, you was cleaning their bodies, and then when they'd eaten all up, they'd maybe rattle their feet, they'd knock on the *skelbases* of their *standings* and you'd feed 'em again. As soon as them horses had finished eating you put another feed in. They were only small quantities, a small saucepan.

By this time the collars would be going on, and then their back

bands and traces. All you did with their *back bands*, you just stood at
' back of the horse and threw 'em, and they went straight over. If
they was for waggons, of course, there was back bands and *belly
bands*. By this time you'd got 'em all *geared*, you brushed up then so
everything was clean, but you didn't use your brush to any muck,
you had a *cowler*, like a massive hoe, and you cowled the muck
straight through the door. Then when you'd got it all, then the brush
come out.

By this time, we all went for breakfast. Some places blew whistles,
some rung a bell and some had a time. Well this place you had a
time, and as you was going in, Foreman'd come out, and all them
fireplaces had a side boiler and they filled up with cold water and he
came out with a boiler tin. They used to hold two or three pints of
water and he put it in a bowl. He got washed, then Wag got washed,
then Thoddy got washed, then Fowth Lad got washed, and if there
was anybody else they got washed at ' back of you.

Then you went in for your breakfast. Foreman was in the armchair
at ' side, Wag had an armchair at end of ' table, then the two lads on
the *form*. You'd fat bacon and dry bread and a pint of tea, a lump of
fat about ' size of both hands, just turning pink on the edge, and
there'd be a piece of lean beef on a Monday morning. When you'd
ate that, you got your bread and then you could butter it, and you
could put some jam on it, or marmalade. Now as soon as you'd
finished that, on the front of the table was the pies, there'd be maybe
a prune pie and a fig pie, and there was a knife laid on top of each
one so you cut your own. Foreman cut his piece and then Wag got
his, and then Third Lad, then Fourth Lad, and then the next pie come
along. You couldn't have any two pieces of one, but you could have
a piece of each. So you got the other pie and you cut your pieces,
each going in seniority, and you always knew when it was figs, 'cause
Foreman'd take his teeth out and put his hat over 'em, put 'em on
table. He had a trilby and put his trilby over ' top of his teeth while
he'd got his fig pie eaten, then he'd put his teeth back in. 'Cause there
was so many seeds in them. Then he'd draw hisself back, knock his
pipe out and you'd gone, and if you hadn't eaten as fast as them,
they were going, and you used to put your hand up, get your lump
of pie and get out ' the house eating it, and straight back into ' stable.

Breakfast was finished shortly after seven, and when the lads got
back into the stable, all the labourers were there for the start of work

at ten past seven. The first job for the lads was to plait up the horses'
tails into a *scotch bob* before taking them out to work. Then all the
blinkered bridles, known as *blinders*, were put on, and if any horse
was to go in a cart, then the britchin and *cart saddle* were then put on,
once the horses had finished eating.

Wag had told us what harness to put on, 'Gear 'em all for waggons'
or, if you were going in a cart, you'd put their collars on before you
went in for breakfast, but you'd never put cart saddles on 'em 'til
they told you, because if you tightened 'em up, they couldn't eat
very well.

Then Foreman'd come in, and he'd say, 'Good morning,' and he'd
walk down giving 'em orders, and then just turned round and
shouted, 'Get 'em out,' and by that, they were all gone and all their
tails were up, and if they weren't up when he come in, he'd come
straight away and he'd plait their tail up in seconds. And sometimes
we'd get just out ' the door and mine had fallen down, and he'd say
'Whoa,' and by that, he'd just shove her up. 'There, you're alright
now,' and we were going to get straw for bullocks.

Well, there was two farms very close to one other, and so I'd to
go down to next door where the waggon was, and at Carr Farm
there'd be Wag with a waggon and Third Lad with a waggon in that
yard, and I'd take my horses into the next yard with a labourer, one
labourer would fork me. You'd go in with wheat straw and bed up.
They'd never been bedded up much on a Sunday and so they'd
maybe want two load. Then we would come back and get a load of
barley straw, and pull in the middle of the yard, and then you just
took the horses off, put ' pole under the waggon, and left the
waggon stood loadened with barley straw, so that the bullocks could
pull at that barley straw when they wanted it. Well by now, the boss,
John Caley, would be coming round and you'd see him talking to
Foreman while they were strawing up. Foreman would be helping,
but he would be talking to him and giving him his orders for the day.

I would then put one horse in a cart, and each yard would get a
cart load of mangels, because mangels weren't considered very good
feed until after Christmas, 'til they'd got mellow. I would have to do
the three yards of bullocks at both places with a cart load of mangels,
get them out of stackyard and throw them on the wheat straw, and
all the bullocks would then just scalp 'em, eat 'em whole. And in one
yard there was always this old bullocky feller and he'd help open

gates for me and get me in. But the other two yards you helped yourself and it was sometimes a bit of a struggle to get that gate open and get that 'oss and cart through before them bullocks come out.

Then the labourers would go to hedges and dykes, and Waggoner and Thod Lad would go to plough, or whatever was wanted doing. Everything had gone by about nine o'clock, everybody'd be out the yard but one old labourer, who was back door man, he had to milk boss's cow. Anyway, by ' time I'd got me mangels done, I'd go and take a load of wuzzels to ' sheep, then I would have to do some odd jobs while dinnertime, while the other two were ploughing.

By ten to twelve, all the horses had come back to the yard, each lad riding his *nearside* horse. But before going into the stable the horses first went into the pond so they could have a drink before eating, and it also washed some of the mud off their legs. Then as they reached the stable door all the blinders were taken off, so they could see more easily in the dark stable.

Then you followed them in, and they'd all gone up their stalls by this time, and your blinders were all hung at the front, and then you fed 'em, a double feed of chaff, a double feed of corn, depending on what work they were doing. But at the top of the work, mainly at spring of ' year when they were working hard, they were eating about a stone and a half of oats each day, each horse; put it in, shook it up with your hand and went for your dinners.

By this time it'd be ten past twelve. Foreman would be in, we'd all march in, nobody washed their hands for their dinners. Tatie, fat bacon, then a pudding: either rice, tapioca, sago, such as that. As soon as you'd eaten it you'd gone, back into ' stable.

When you'd got back into ' stable, roughly about half past twelve, the first thing they did was feed ' horses again. Then I'd to get a *caff sheet* and get Waggoner some chaff and push a sheetful in his *caff hole*. Third Lad, his caff hole was near so he could get his quite easy. Then Waggoner and Third Lad had to go up ' grainery and get their oats, and they used to lay their bag down and fill it with their arm. Then they put their corn in their corn bins. Then you wanted some mangels, and we'd to go with a bag to ' mangel pie, or wuzzel pie as we called it. You had a knife and a *drag*, and the mangels was cleaned and put in a bag, and each horse was going to eat two mangels, which after Christmas was going to be four mangels. But they were always cleaned for them, just all the soil knocked off 'em.

By the time these jobs were done it was nearly one o'clock, and the labourers would join the lads in the stable, sitting down for a few minutes on the bales along the back wall until the foreman arrived. He might then ask them how they were getting on with their work, or give instructions for the rest of the day, and often Ron would get a pair of horses ready to go ploughing with the other horselads.

> One o'clock, same procedure: 'Get 'em out!' Now we put the blinders on, they were put on inside the stalls, and each horse was let go and you followed with your other horse, and they all stood waiting for the other horse coming up to them. There was nothing to stop them going either way, but they never moved, they just stood. And of course, when everybody was going, there was five pair went out, and five horses would be stood out there waiting for us going with the other five, and you just tied your *farside horse* to your nearside horse, put one hand on your farside *hame* and jumped, and you was sat on their backs and off you went, and you all rode in formation. If Foreman was going with you, he rode the first pair, then Waggoner, then the Third Lad, then Fourth Lad and any labourers go after.

During the winter, work in the fields continued until dusk, but if the journey back to the stable required the horses to go down the road, the waggoner needed to make sure they had finished work and had got back into the yard before darkness fell.

> This would be nearly half past four, and, same procedure; into ' pond, give 'em a drink, come back in, strip 'em, let their tails down, give 'em a feed. We'd plenty of time. We'd brush; if there was any muck in ' stable, you brushed it out.
>
> Then we went in for our teas, just gone five o'clock. Same procedure, fat bacon. Now if over the course of the time you were getting a bit nearer the meat, there wasn't quite so much fat, because those pigs weighed forty stone. They used to get an old sow what had had a litter on 'em and the pig used to be nothing but skin and bones, and then they used to feed it up and make the meat, it was quite sweet. And then you got your dry bread again and a pint of tea, then your bread and butter, or margarine, 'cause rationing was still on at this time. Although the war was over, it was still rationed, and bread had now become rationed which it had never been in the wartime. Then out come the pies again. Well by this time, they were maybe different. There would be curd cheesecake, or ground rice cheesecake

and date pie. There was no fruit trees on this farm at all, not at Foreman's house, so all your winter pies were dried stuff. The prunes, they weren't stoned and they weren't soaked much, they were just thrown in, and by hell, you hadn't half to champ. A lot of these pies were on these tin plates with a big flange on, and there was a hell of a lot of pastry, but not very much inside. If you cut a lump of pie you might have two prunes in there, or three prunes, and it wasn't short crust pastry, it was quite solid. You wanted something to drink to get it down you, and that was your last meal of the day, and as soon as you'd got your tea, you'd gone, back into ' stable.

After tea, there was no hurry to get anything done, so when the horses were turned out into the fold yard to have a drink, the lads were happy to leave them to play and roll. While the horses were out in the yard, fresh straw was put in the stalls for the night, and a feed put in the cribs for when the horses were called to come back into the stable.

We bedded 'em up, and there was no *heck meat* at all, there was nothing. All they got was oats and wheat chaff, and when they'd eaten that we started to feed 'em again and then we cleaned 'em a bit. If they were dry, depending what we'd been doing, wisp of straw, wisp 'em down their shoulders in particular. You kept feeding them, every time they finished eating you was feeding them, and then you left 'em with a feed, and then you give 'em the mangels, two mangels up to Christmas, and four after Christmas.

By half past six or seven all the work in the stable was finished, the horses were fed and bedded up, and the lads were then free for the evening.

Then you could do as you liked, but there was nothing to do; there wasn't a pub. It was three miles from anywhere at Flinton, you went three mile to Sproatley, nearly three mile to Aldborough. Then if you went sideways you got to a little hamlet called Tansterne. There was nothing there whatsoever, and if you went the other way you went to Humbleton and they hadn't even a pub at Humbleton. But we'd no money much, so we just stopped in ' stable 'til you thought you wanted to go to bed.

This daily routine continued throughout the week, though on Saturdays everyone stopped work earlier, but the horselads still had to do their horses after work. On Sunday, however, the horses were not at work, but they still needed to be fed, and it was still the lads' responsibility to feed them.

> Now Sunday morning, we always said we could have a lay in. So we didn't get up while six on Sunday, and went in for our breakfast at seven the same. But by this time we hadn't done much in ' stable, time was our own. So we come back in and we brushed up and we always put the beds down on a Sunday, and it was surprising how many used to lay down. We used to comb the manes and comb their tails out on a Sunday morning, and we'd turn 'em out again after breakfast so they could play a bit. We could bed 'em up and sweep out, we was in no hurry because we wasn't going nowhere. We used to leave 'em in there for about an hour and they used to get salt from one another, all biting for salt, then get 'em back in, and the beds were all down of course, and they'd lay down.

Once the horses were back in the stable, the lads were then free until the horses needed to be fed again at dinnertime. One of the ways the lads passed the time was to go rabbiting.

> We all used to go ferreting, but you wasn't allowed to go in one place. You could go anywhere on the farm but you weren't sup- posed to go on this place and that's where the rabbits were. Of course we used to go there when they went to church!

Sunday was the lads' only chance to have a rest, and on cold days in winter they made the most of an empty standing in the stable to get some sleep.

> In winter time, they didn't allus have a full contingent of horses in. There was twelve horses, and there was stalls for ten at Flinton, and at Manor Farm they used to have a horse yard, which was very unusual; very rare they did that in Holderness. They used to put these horses in there for winter and they never got 'em out. They just used to stop in there; you didn't need 'em. So there was a stall left in the stable at Carr Farm and we used to fill it up with straw. And we used to get a bag and fill it wi' straw, and get in this bag and go to sleep 'til dinnertime. 'Cause you had to be there, you had to do your horses on a Sunday morning and do 'em on Sunday night as well, so you were all kicking yer heels.
>
> We had to give 'em a feed at dinnertime, and then your time was

your own 'til half past three. 'Cause you had to have your best clothes on when you went in for your tea, and on a teatime, then you ought to see a spread on the table! There'd be about seven lots of cakes and pies, and Wag used to say, 'Eat a bit of each, else they won't fetch so much out next week!'

And on a Saturday night and a Sunday night they left us supper out. There was always a teapot there with a tea cozy on and a couple of lumps of pie each.

Throughout the winter and most of the spring the horselads' routine remained the same, until the horses were finally turned out to grass in May. Until then the Sunday break was the only chance they had to have some rest before the foreman's call on Monday morning heralded another week of work.

CHAPTER 5

Winter at Carr Farm

BY the end of Martinmas week when the lads arrived at their new place of work, it was already the beginning of December, the start of the farming winter. Traditionally in Holderness, much of the ploughing would already have been completed, the plough teams having been out at work as soon as harvest was over. But the wartime introduction of potatoes into the rotation rather upset the traditional pattern, as the potatoes needed to be lifted at the time of year when the ploughs would otherwise have been out in the fields. So in Ron's time there was often still some ploughing left to be done at Martinmas.

> When everything was completely horses in this district, there was no taties. So the minute they finished harvesting and they'd got all the *rakings* off, they would be ploughing, because they was going to sow winter wheat and beans. They used to like to sow all the winter wheat in the last week of October and the first week of November. They'd maybe be growing hundred acres of wheat at Flinton, and when they used to go to plough all with horses, they used to turn out about seven ploughs. They'd be ploughing seven acres a day, so after a fortnight they'd have it done.
>
> But if they got nearly ploughed up and it was nearly Mart'mas, some on 'em would stop you ploughing so that it left twenty acres for the new lads coming in. Their first job was to go to plough so he could see what they were made on. But once they started growing taties, they never got ploughed up by November.

Generally the horselads all ploughed together in the same field. On the neighbouring Wolds, the horses and ploughs all followed one another on the same piece of ground, but in Holderness each lad ploughed his own piece. The exception to this rule was when setting out the field at the beginning of the process. Setting the field out was necessary because on the conventional or one-way ploughs used on most farms, the mouldboard turned all the furrows to the right. So if you were to plough a furrow at one side of a field, turning the furrow towards the hedge, when you reached the other end and turned round, the mouldboard would turn the next furrow towards the centre of the field, rather than laying it against the first one. To overcome this problem the field was marked out into sections, or lands, and a ridge, or *rigg*, ploughed in the middle of each land. The rigg was formed by turning the first two furrows away from each other, less deep than normal ploughing, leaving a shallow furrow. The two furrow slices were then turned back towards each other, but the plough was set more deeply so it also cut into the undisturbed ground, and these deeper furrow slices formed the ridge. The ploughman then continued to lay the subsequent furrows inwards towards the ridge, turning to the right at each end of the field.

Once two adjacent riggs were ploughed, leaving a narrower piece of unploughed land in between, the horseman then turned his horses to the left at each end, so as to lay one furrow towards his own rigg, and on the return journey turning the furrow towards the adjacent rigg. This was known as 'arving out', *arve* or *orve* being the command used in Yorkshire to turn to the left. As the strip of unploughed land became narrow, the horseman checked that it was parallel, and adjusted the plough to correct any inaccuracy, so that all the land was ploughed. The final furrows were cut more shallowly so as not to leave too great a mark on the land for the subsequent operations. In order to make sure that the ploughs were ploughing the whole length of the field, it was important that the riggs were parallel to each other and that the furrows were straight, so it was the waggoner and foreman who set out the fields.

Foreman would measure while Wagg'ner was ploughing a mark out, and he would put one stick on the headland, another stick ten yards in, and another stick in the middle of the field, because you're always straight with two marks, it's the third stick that tells you just how straight you are. And as he measured across thirty yards, and put his first stick in, the waggoner had cut his first mark out, opening it out, and then he would go back down the same mark, opening it out the other way. By this time Foreman had got his sticks set up for the next one. The waggoner would then move over, and plough a mark and open it out again, and the foreman would keep walking and measuring up; and if the field was twenty acres they would set the riggs in the complete field at thirty yards apart. Probably we wasn't there that day, we'd be doing something else, and it would be nothing for them to go into two or three fields and set the riggs up in all the fields. Because when we came, we'd nothing to do then but plough. You closed all these [riggs] up, and you each had one apiece so you was thirty yards apart, and if they was so minded Foreman would come, and Wagg'ner and Third Lad and Fourth Lad, and if there was some horses left they would have one of the young married labourers who'd been engaged to go with horses and he would plough, so if there were ten horses, there would be five pairs ploughing.

We would all plough, close them [the riggs] up, then you would plough ten yards either side of your rigg, and the other man would have gone ten yards each side of his rigg, leaving ten yards unploughed in the middle. When you was going round your rigg, you were going *gee* round, and 'gee back' means going right, and you said orve or whorve to go left. So when you'd each got your ten yards round, that left you ten yards. So you each went into your left-hand piece to plough orve round, but you didn't finish your *forrs*, they were all left. Then your Third Lad would finish all forrs, and we used to sole out; when they'd got it all ploughed you used to come down the last forr, just taking the bottom out.

Even when everything had been ploughed and the wheat and the beans were sown, there was still more ploughing to be done over the winter. On the land which would be sown to roots next spring, the furrows that had already been ploughed were either turned back to their original position, or the whole field would be ploughed again at ninety degrees to the original ploughing, quarting as it was known. The point of all this ploughing, particularly on heavy land, was to allow the frost into the lumpy clods left on the quarted land, to break the soil down into a fine enough tilth for sowing the small-seeded root crops.

In Holderness the use of three horses on a single-furrow plough was widespread, because of the hard-working clay soil. Sometimes the three horses were yoked abreast, but yoking *bodkin* fashion, with two horses in the furrow and one on the land, was also common. However, Ron's experience was of ploughing with a pair, much of the ploughing by this time having been taken over by the tractors. In a wet time, however, the tractors were put away and the tractor drivers also went to plough with a pair of horses. When it was wet, or if the harvest had been wet, as it was in 1946, the smaller land wheel of the horse plough, which helps maintain the depth of the ploughing and increases the plough's stability, was dispensed with.

> We only ploughed with one wheel: you never had two wheels on. The first thing they did was get that wheel off, because if you was ploughing wi' two wheels on and it had been a wet harvest, them waggons had cut to hell out on it, and when you got near to ' gate you couldn't see where the hell they'd been. You imagine ploughing down there; if your little wheel got down one rut, it'd be dropping in and getting fast. With them ploughs, oh hell, you'd only them one wheels on, and as you were going round [on the headland], if it caught you, you were thrown. With two wheels on, they can't come down, but with one wheel on, one set of *hales* catches you here and the other there, and spins you down.

The ploughs used at Carr Farm, with wooden stilts and a wooden beam, were made by the Holderness plough company. Even though by the early twentieth century there were big factories, such as Ransome's of Ipswich and Howard's of Bedford, making huge numbers of iron-framed ploughs, there were still many smaller firms making ploughs which were particularly suited to the local conditions. Although the wearing parts of all ploughs, the *sock* (or share), the knife *coulter*, the *skeef* (or disc coulter), and the mouldboard were all made of iron, the wooden beam and hales made these ploughs significantly lighter than an iron plough.

> They were all wooden ploughs, Holderness ploughs, made in Burton Pidsea village. That was one foundry, and the other foundry was at Ellerby, Grassby's, they used to make wooden ploughs as well. They all come with a knife in the sock. All the socks round here had a knife on 'em, it come to two or three inches high. And all them wooden ploughs had a place for a coulter and they had a place for a skeef as well, so you could have all three if you wanted.

With the ploughing in autumn, quarting in winter and ploughing fallows* in the summer, there was nearly always some ploughing to be done, so the foremen would make the best of any opportunity

---

* The last bare fallows at Carr Farm were in 1947, and were ploughed with tractors.

66

to complete the work, even sending the lads to plough for an hour in a field near the farm buildings until they were wanted for something else. Besides working in the fields, in the winter there was always work to do around the yard first thing in the morning, much of it involving feeding and bedding up the bullocks. When the bullocks were first brought into the yards in early winter and the muck was not very deep, the straw was often moved on a waggon pulled by a single horse. Most farm waggons throughout the country have shafts, whereas the Yorkshire waggons, with the exception of the smaller Moors and Dales waggons, have a pole for two horses. When two horses are put in a pole waggon, the chains from the end of the pole to the horses' hames allow it to be steered, but when one horse was used without a pole, there was no way of steering. Nevertheless, pulling a waggon *swing* as it was called, was common in Ron's experience of Holderness farms.

> When we first went, all the waggons were pulled swing. We didn't always put ' pole in, we used to just put the chains on the hooks and pull 'em all over yard; just one horse pulling loose, no brakes, no pole, no nothing.

As the winter progressed and the muck in the fold yards got higher, the waggon wheels sank into the muck, requiring a pair of horses to pull the waggon, so the pole was put back in place. With the constant movement of waggons around the stackyard and despite each load being scraped down with a fork before the waggon moved away from the straw stack, some bits would fall off, and when combined with soil from the waggons' wheels and chaff blown about on threshing days, the stackyard soon got mucky. As soon as this happened, the foreman sent everyone to go and clean it up.

> We were always shovelling, cleaning *blather* up. Everybody would be there; two men would have cowlers and they'd use them like hoes, and bring the blather into rows from both sides. Then you'd go down between with a cart or a rulley, but mainly, this is where we would use carts, and they'd shovel up into carts. There'd be two men filling, the man with the cart and another man. Then at ' back of that there'd be men with brushes, and we'd go round the complete farmyard, and we could walk down to ' stable in our shoes on a Sunday.

While the horselads were going between the bullock yards and the stackyard with waggon loads of straw every morning, the labourers were busy feeding the bullocks in the three different yards, the bullocks being divided into groups depending on their age.

> In those days bullocks weren't killed 'til they were nearly four. So the first year, what we'd call good calves, they were fed well; oh hell, they weren't half fed. Then they were turned out to grass. But the second year they come up, they nearly went out thinner than when they come in. All they got was *chaff* and wuzzels all on the floor, barley straw to eat of course; there was always loads of barley straw pulled in, or oat straw. Then the third year the bullocks come up, then they were fed! Each tum'ril got a basket full to start with, ' get 'em eating, then you kept going round, and then when they'd finished they used to be groaning and moaning, and they'd all flop down and they'd lay there, and woe betide anybody who stirred 'em all up! And if they started to lick each other, that was what they used to reckon was a good sign they were doing well.

In each of the yards there was a big trough, about six foot square and three foot six deep, for the cattle to drink from. Except for first thing in the morning, the horses usually drank from the pond after they had been at work, but before piped water arrived, the cattle relied on the ponds in the summer when they were out at grass, and the rain water which filled these troughs in winter.

> Water was a bit of a problem, and all those buildings were all connected up, one gutter would run into another and it'd all assemble at one point and then go into a big tank, and that was what water there was on the places. There was no taps, and all that water usually just about kept enough water going, and if it didn't, they'd water carts. Every house had rainwater barrels and rainwater tanks. Although there was one tap in the wash house, there wasn't none in the house; all the water was got out of these rain tanks and put in the side boiler, and you was always getting washed in rain water.

During the winter a regular job was threshing the corn, beans and peas, so that most of the stacks would be threshed by the end of spring, but even during the busy autumn period of ploughing, harrowing and drilling the corn, a start had been made on the threshing.

> They'd had a few days threshing, they had to thresh for straw to *hap* these stacks down and also for seed to sow, and probably for money as well. And if we were going to have taties we stacked a day or

two's threshing of corn down where we were going to have taties,
so we hadn't to cart the straw.

Once winter arrived, however, instead of threshing just for a day
or two at one farm before moving onto the next one, the threshing
machines sometimes stayed for a week or more at some of the bigger
farms which had many days of threshing to do.

On this particular place we'd forty days threshing.* The machine was
owned by the company, and they were threshing every day it was
possible from just after harvest, and at one time you got more
money for your wheat if you threshed it in June, and so they'd be
threshing right the way through.

At Carr Farm there was enough staff to make up the gang necessary to run the threshing machine, with the foreman, three horselads,
two tractor drivers and four labourers, but at times some of them
would go to Pasture House to make up a threshing gang.

He had another farm as well, had John Caley, Pasture House, which
was about a mile and a half away. And when they was going to
thresh, so many labourers used to go from Flinton, 'cause you want
eleven men to do a day's threshing; one lad carried *caff*, one lad
carried *pulls*, there was two men on the machine, one feeder and
one band cutter, two men on the corn stack, two men on the straw
stack, two men carrying corn.

The jobs of carrying the chaff, which would later be fed to the
horses and cattle, and carrying the broken straw and leaf, known as
pulls, were always lads' jobs. Unlike a combine, a threshing machine
separates the pulls from the good long straw, which falls from one
end of the machine, either into a trusser which makes low density
bales, known as trusses or *bottles*, or onto an elevator which carries
the loose straw up to the straw stack. The pulls, meanwhile, fall to
the ground in between the threshing machine and the elevator, and
are raked out onto a sheet, which is then carried into the fold yard to
be used as bedding.

With this being such a wet harvest, when we come to thresh, the
straw broke up and there was as many pulls as straw, and you were
getting bunged up. You left your rake against the machine and the

* This is probably somewhat exaggerated.

corn carriers would come and just rake it out. But by ' time dinner-time come, your sheet was on top of all these pulls, and you was raking 'em uphill. So when you'd had your dinner you used to have to go back and carry all these pulls away in your dinnertime, so you could get caught up. Then by ' time night came you was just about bunged up again, but of course that was the end of the day's thresh-ing. Ooh, it was bad for breaking up, and if it got too bad, we could go to the elevator wheel and just give the elevator wheel a push, and the belt used to fly off and you'd shout, 'Belt off,' and by ' time they'd messed about putting the belt on, it give you a chance to get a sheetful or two took away.

The wet weather during the harvest of 1946 had also forced changes in the way the barley was stacked. Instead of making big stacks, tall narrow stacks which dried out more quickly were built.

In a bad time you'd be leading it damp and wet, you'd no alterna-tive; they used to make stacks just a load wide and as high as they could get 'em to go. Usually barley was bad to stack, very slippery, but when it was on the wet side it didn't slip, so they'd be about three loads high and a load wide, but they went in length instead. They just kept stacking 'em to get 'em in; made it a bit more difficult, but it's ' only way they could get 'em. Whereas a wheat stack, your bottom was about nine by five [yards], eight by five, something like that, and you could lead wheat when it's pouring down with rain. Ooh, and if it was a bad time we led it in the rain and of course you only wanted about the first three *shavs* and your knees were wet through.

And these stacks, as long as you didn't want to thresh 'em while about spring, when you opened 'em out there was just one white coat of mould, and you could go with a sledgehammer, it was that hard; the straw was just like cardboard. It wasn't very good straw, but it made bedding straw. And you put these sheaves through the threshing machine and all the mould had gone, but ooh, it was all in the air, especially when they pulled these machines between two stacks, and the caff carriers were in between two stacks and the machine, and all this dust was getting up your nose. It wasn't very pleasant for nobody, everybody was breathing in this dust.

While the jobs of carrying the pulls and chaff were lads' jobs, to become a waggoner in the East Riding, the yearly agreement required you to carry the sacks of corn. Most of the carrying occurred on threshing days.

The corn, after it had been separated from the straw in the threshing machine, came out of spouts at the opposite end of the machine from the straw, and fell into hessian sacks. These were then lifted up on a winding up barrow, so the man could get under it and carry it across his shoulders, up the granary steps and into the granary. Sometimes the corn was left in the sacks in the granary, but if the corn was for home consumption the sacks were often emptied, or *shot*, into a heap on the granary floor. Although the stackyards were laid out to be convenient for the granary, the position of the stacks also needed to allow enough room for the threshing machine and the straw elevator. On the bigger farms with many stacks, there was quite a distance for the waggoner to carry the corn, a task demanding both strength and stamina.

> If you was Wagg'ner you'd have to carry corn, and the biggest weights was twenty stone; beans was in twenty stone, peas was in nineteen stone, wheat was in eighteen stone, barley was in sixteen stone and oats was in twelve stone. And it was the same bags, but with each vary of corn weighed different, you could get 'em in the bag. And Flinton was a hard-carrying place, the longest carry was nearly fifty yards and then up the steps and ' grainery – oh hell, ' grainery was such a long grainery.

If the grain was going to be sold off the farm, in addition to carrying the bags, each one needed to be weighed and the top tied. In earlier times, the sacks of corn were stacked on a pole waggon, and delivered by the waggoner to the mills in Hull, one of which was owned by Joseph Ranks, who first started his business from a windmill in Sproatley. The round trip to Hull was fourteen miles, but in Ron's day, this opportunity for the waggoner to make an impression with his fine horses on the way to town had passed, and instead the corn was taken away by motor lorry.

The corn was still collected in sacks, and on a hard threshing day weighing the bags of grain was an extra job which was not always done very carefully, according to John Caley, who would insist the lads take more care.

> When they were threshing fast, they weren't too keen on weighing it. As long as they put them on the scales and it went down, they'd tie them, and he used to come round and say, 'Bacca weight. If you

go for your 'bacca to the shop, you don't get more than your
weight, do you? Make sure them bags are right! Bacca weight,' he
used to say, 'You're the lads that made Joseph Ranks.'

Lifting and carrying were an everyday part of work on a farm in
the 1940s, and although much of the hard work was accomplished by
skill as much as sheer strength, being strong was still something to be
prized, so particularly on days where the men and lads were working
together, demonstrations of feats of strength were not uncommon.
Foremost among those demonstrating their prowess was the foreman, George Gibson.

> Foreman and Waggoner used to carry corn at Flinton, and one day
> we was taking a waggon load down to ' next place and I was
> handing off, and one of ' labourers said, 'You've to carry ' last bag
> up, Ron,' and Foreman says, 'No, he's got that wrong. Whoever
> carries ' last bag has to carry you up.' Well, nowt no more was said.
>
> Well, there was three labouring fellers, they took bag for bag and
> when it come to ' last bag it was Foreman's. So he got it; he says,
> 'Jump on!' And I got on, just straddle legged like a horse, and up the
> steps he went and he said, 'When I was young, I was strong,' and I
> thought, 'By hell, Mister, I bet you was!'

Other instances of George Gibson not only demonstrating how
strong he was, but also showing how he wanted things done,
occurred when leading straw from the stackyard.

> When you used to pull these waggons swing, up to these straw
> stacks, he'd say, 'That isn't where I want it,' and he'd just go under-
> neath and get his back under and lift, and when you lifted, you had
> to lift a bit more, because the body wasn't bolted tight to the wheels.
> The body lifted up three or four inches before you lift the wheels and
> the axle, and he just used to walk sideways and put it down. 'That's
> where I want it,' he used to say, and I've seen him do that more than
> once. They weighed twenty-five hundredweight, did a waggon. He
> wasn't showing off, he was showing you where he wanted it.

It was not just George Gibson who used to take part in such antics;
some of the labourers would also join in, particularly on wet days
when they were in the granary. One trick was to tie a handkerchief
to a four-stone weight and lift it with their teeth, and George Gibson
would lift one of these weights in each hand and hold them out to
the sides and then out in front of him, as if doing his exercises. Some

of the labourers would also demonstrate their agility by swinging from beams.

> They'd throw ropes over these beams. They'd get hold of them and twizzle round and round, and take their cap off and as they were going round and they'd pick their cap up with their feet, and I've seen these fellers sixty-odd do it, jump in there and twizzle round and round and get their caps and go round. They'd done it all their lives, done it when they were lads and they'd carried on doing it.

The main protagonist in these acrobatics was Ted Simpson, known as 'Codge,' who had started work as a horselad at Carr Farm in 1900, when Charlie Buck was the foreman.

> He would tell the tale, would Ted, that he could lift hisself in a scuttle. A scuttle's sort of a tin bin with two handles on, and we've seen Ted get in this scuttle and get one hand on each handle, and he would jump, and of course the scuttle would lift up.
> 'You nearly did it, Ted.'
> 'Aye, I know,' and he would jump again and it would lift.
> 'By hell, you nearly did it, Ted.'
> 'Aye,' he said, 'I used to be able to do it when I was young,' and he was well known in the district; they used to have him doing this in the pub yards: Ted in the scuttle.
> Poor old Ted, he was nearly sixty-five and he used come for a 'oss, put a 'oss in a cart, and he used to say, 'You couldn't help me to get [a bag of wheat],' and they expected him to get eighteen stone of wheat on his back, and walk down ' grainery steps; and it's much harder to walk down grainery steps than to walk up 'em, and they used to put this eighteen stone of wheat in the cart, and, 'Aye why, if you can get me another one?' So I'd get him another one and put it in the cart, and off he used to go wi' chickens, and he used to take it to these different bins; we had chickens all over the place, and then at night he used to bring me a double-yoked egg, and he used to say, 'Here you are, look, you can get somebody to cook that for you for your supper,' because we never got any supper on these farms, only at ' weekend. Poor old Ted, we used to feel a bit sorry for him.

When the workforce was congregated in the granary on rainy days, the job they were supposed to be doing was sorting out the piles of empty sacks which had collected over the months.

These bags were hired. Our bags was mainly Chisholm, Fox and Garner; London, Liverpool and Hull; them was the three main ports in England, and when you hired these bags they was three ha'pence the first week, then a penny, and eventually I think they went down to a ha'penny. So each bag would hold different weights, and they were all the same money. How the hell they kept tag on these bags I don't know, because lorries would come with seed corn, they were all in the same bags, and then there'd be lorries taking loads away; you'd put corn on ' station, and nowhere did you do any writing. And at ' end of ' year he'd come along and he'd say, 'We're a bag or two short. Can you have a rake round?' And we'd go round and the first thing was you went on the tractor seats, there'd be two or three bags on each seat, and there might be one in Foreman's house as a hearth rug, another one at back door and you'd go round raking up round. Then he'd say, 'We're still about twenty short.' So they'd go and hire twenty and put 'em in, and make it right, and then set off again.

And a lot of these pubs in the villages, they were the depots for these bags, and all these bags were done in twenties, nineteen inside, and one what it was in. They were all folded in a particular way and these depots used to empty 'em out, and you were supposed to *shak* 'em out and clean 'em all out, and you took 'em to and fro.

And then the railway bags, London, North Eastern Railway, you went to ' railway station for them. They was all ' similar price. And sometimes the lorry would come and you'd fetch a load of maybe five ton and you'd give him same quantity of bags back empty, but we never writ nowt down. And these bags, if you lost 'em, were charged at something like seven and sixpence. And on a wet day they'd say, 'Up grainery!' and you went and you started sorting all these bags out. You found all these Ranks's bags, and you'd find somebody else's and somebody else's, and the money that was in these heaps, half a crown a bag and three shillings a bag, and it was all stamped on 'em. Then eventually you'd tie 'em all up and they would take 'em back, and ' course, they'd get their money on 'em.

The other wet-weather job in the granary was bagging up the heaps of corn shot onto the floor on threshing days. The quickest method of filling the sacks was to lay the sack down and pull the corn in with your forearm.

At one point the grainery was full, nearly from one end to the other with barley, and as they carried it up they'd shot it all; and then they wanted it all bagging up. They sent me and one of these labouring lads to bag it up and the first thing Foreman did was collect the shovels up and take 'em away.

We each had a bag and we scooped it in with our forearms, and it's all very well 'til you want the last bit in. But anyway, we didn't weigh 'em as we did 'em, we just filled 'em and filled 'em, and I went down and fetched a shovel back. Of course when we started to weigh it, that's when you wanted your shovel, just to top 'em up, or a tin to keep the weight right. And the first thing he did when he sent you to work up grainery, was open one door at one end and one door at other end so you didn't get too warm! And the wind would come in at one end and out the other, and he always did this!

Another job kept for a wet day was mucking out the loose boxes housing the young unbroken horses, which had been brought in from grass about Martinmas time. Except for putting straw and mangels into the boxes, the horselads had nothing to do with the young horses, which were fed by the foreman or one of the labourers.

When they had these boxes with young horses in, there was maybe two or three in a box, and they never mucked 'em out. All we did was bed 'em up. When we come round with straw for bullocks we used to pull tight up to ' doors and throw this straw in, and throw it all over these 'osses, and by hell, didn't they used to kick and gallop, and they hadn't much room, and they was bedding up theirselves.

They'd fed 'em earlier on in the morning, and then when we come round with wuzzels we used to just throw these wuzzels through the top of the door and if it hit 'em, it hit 'em, and if it landed on the floor it meant they ate 'em off the floor, and then again they fed 'em at night; nothing to do with us whatsoever.

The only time we mucked these horses out was when it was a wet day, and if it was a dry winter you found out that the straw was getting up as high as the top door, and what a struggle it was to get in to feed 'em. And if it was wet, pouring down with rain, 'Muck the 'osses out!'

They'd never been touched, they'd no helters on, nothing. So we'd open the doors and prick 'em with a fork and they'd jump out into the bullock yards, and then while they were out we'd muck 'em out, but we left six inches of muck in. They'd been used to heat under 'em, so that six inches kept the heat under these horses. When we'd got that box mucked out you'd fetch a jack straw full of straw, bed 'em up, go and round 'em up and run 'em back in the box, and then have a go at the next box.

And then the days it rained and there was no 'osses to muck out, it was up ' grainery, and they was always cob-webbing, brushing cobwebs out, and somebody would be outside cleaning all the spouts out with a bucket and a little hoe head. And if the Brigadier came round, as landlord, the first thing he did was look up there and point to the spouts and say, 'Those spouts want cleaning out.' Very particular about his spouts: that's about all he knew was spouts, 'cause they was professional soldiers, generations of professional soldiers.

Rainy days were also an ideal opportunity to send the horses to the blacksmith. Besides shoeing horses, the blacksmith's work also included repairing implements, and in Holderness one of their main tasks during the winter was to hammer-weld new points on the harrow tines, which wore quickly in the heavy soils.

We used to run two blacksmiths, Humbleton blacksmiths and Aldborough blacksmiths, and these blacksmiths shod their horses and other people's as well. But they also *laid* all the harrows, straight after when we'd all sown up in the autumn. You'd use the harrows at spring and then you'd use them at the autumn, and then whatever harrows wanted laying we would put 'em in a rulley, and he'd take 'em to wherever they were going. They'd take all these harrows off and they were all stood in the blacksmith's yard. How the hell he knew which they all was, I don't know, there was harrows after harrows. They'd be stood out from the walls, all stood on their end, different people's, right the way round, and that was his bread and butter. He'd be laying them when you went in and took these 'osses to shoe 'em. Some were shod all round, some were only shod on their front feet. Pasture House 'osses were never shod at all, it was a bit lighter was their land, but Flinton 'osses were all shod all round.

But if it was a wet day and spring was coming, Foreman'd come into ' stable and look at 'em all and he'd say, 'Right, blacksmiths!' He'd send somebody to Aldborough with some and somebody to Humbleton with some. You'd set off, and it's flat land and all the hedges were all cut nicely about breast high; you could see miles, and you'd see somebody coming from somewhere, and if you wanted to make the day of it, you held yours back, and if you wanted to get on, you kicked 'em on to get in front on him. When you got there, of course, you'd somebody to talk to while they were shoeing 'em, and I remember going to Humbleton this time and he says, 'Are you cold? Do you want warming up?'

'Yeah.'

'Here you are then, look,' and he give me the box and he said, 'Take them shoes off!'

So he showed me how to take one off, and he went back to laying harrows. And when they laid these harrows they hung 'em up with chains in the blacksmith's shop against the forge and they took 'em out, ' tooth at a time, and then they laid 'em and put 'em back. Some were called ducksfoot harrows and they were just like a duck's foot and they had a little edges on to cut into the land. Then there was chisel harrows, they were *gibb* harrows. They were straight, just like a chisel, but by hell, they cut your boots in two when you were coming back from the blacksmith's shop; hell, they were sharp. The first row were straight-toothed, and three rows of gibb harrows, and a three 'oss set had three harrows, and you could just harrow twelve acres a day with them if you kept going, and a four 'oss set, of course, you could do sixteen. Four acres a harrow, they used to reckon.

Anyway this particular time he was late laying it, 'cause that's what they used to do all winter, and they used to go well into the night, and of course, if they was in the village they used to get half the old people who wanted to warm up and had nowt to do; they all used to sit round in ' blacksmith's shop and talk to him. Anyway he showed me how to take these shoes off, and then I'd been a few times, he says, 'Do you want a go at trimming one up?' And he only gave me front feet, he give me a rasp and showed me what to do. So every time I went I just used to go in there and get the box and take the hoof off ready, so I was a friend of his then. 'Cause they didn't like shoeing them horses – hell no, none of them. And on a wet day, they knew what was going to happen and they got up early and got the fire on. And when I got there, I'd be maybe from three miles away, there was a pair coming out, he'd shod them, and before I'd come out there was some more. All day long when it was wet they were coming from all over.

The newly laid harrows were brought home in the early spring, because as soon as the land was dry enough, the horses were out

harrowing down the land in preparation for drilling. With the freshly
sharpened points on the tines, however, the harrows would dig into
the ground more than when blunt, sometimes making it impossible
for the horses, or even a tractor to pull. To remedy this, often one or
two of the tines was turned to face backwards, so the harrow would
not sink so far into the ground, or some horselads would carry a small
baby milk tin to put over a harrow tooth to do the same job.

In 1947, however, the spring work was put on hold, because at the
end of January the weather changed dramatically, heralding the start
of one of the longest cold periods of the century.

The last Sunday in January, Wag comes in. By, it was bitter cold,
freezing; and he said, 'Come, we'll have them 'osses done!' We was
sat in ' house with Foreman. So we goes into ' stable and does all
these 'osses up, and you had to have them all done on a Sunday so
that you got changed and sat down for your Sunday night tea in
your best suit, somewhere round half past four. So everything was
done by five o'clock so Missis could get all washed up, and she'd
nothing to do for the rest of the Sunday evening. Anyway, we comes
in. They had their teas and Wag said, 'We're going out, Foreman.
You'll give 'em their wuzzels?'

And Foreman says, 'Where are you lot going, then?' Well, one
was going to Hull and one was going to Sproatley, and he said to
me, 'You're stopping here, they won't come back tonight,' and he
says to them, 'You ought not to go, 'cause you aren't going to get
back tonight,' and there was nothing to be seen outside.

'Aye well, we're going,' and off they went.

So he says, 'You get sat again' fire with me!' We used to play cards
and things at night, and the last thing he did before he went to bed
was bring a shovel inside, and next morning when we got up, the
snow was up just below the bedroom windows, drifting right across
the road. But when we come out the back door there wasn't much
snow and so he says, 'You go into stable and I'll come and give you a
hand.' So he come with me and we did the five each, and them two
didn't appear, and then eventually all ' labourers come into ' stable,
and ' course, with it being Monday morning, there was no wuzzels
up at the farm, they was all at the pie, there was nothing prepared,
so we'd to go across these fields and get some wuzzels. So he says,
'We'll have three 'osses in each waggon, you get Wag's 'osses,' and
one of these labouring lads who went with 'osses, he had another

three, and neither on us had ever driven three horses before [as a *unicorn*]. I had Wag's horses, and I'd never had them before either, and all the lot of us, everybody went. We cut across the first grass field, and all the labourers had to dig all the gates out, and then dig the road out. And when you got in some of them fields there was no snow in 'em at all, it had all blown. It was in all the gates and all the roads; the roads east and west were worst.

When we get to the wuzzel pies, they'd pick axes to get the soil off, and when we got inside, the wuzzels were frozen. They had to get 'em down with pick axes and the wuzzels all stuck to the straw, and you could stand underneath them, and what a job it was to get the soil off! We'd loaden all the three waggons up and take 'em home for bullocks. Then they'd start filling [loose] boxes and then we'd loaden all the three waggons up and take 'em home for next morning.

Then ' next day, when they came in the stable, he says to these labourers, 'Anyone who can get snow digging, do so,' 'cause he didn't want 'em on their books; there was nowt to do. So anyone who could get round the council and go snow digging, that was rid of them.

Anyway, by about Wednesday, Third Lad comes back and by Thursday, Wagg'ner comes back, and there was no buses, 'cause it didn't stop snowing. The buses had got as far as a village called Bilton which was best part of five miles away, and every time they got the snow dug out, next morning it blew it all in. And we never got another thing for six weeks, anything with a motor. 'Osses used to come through; the coal men eventually come through after about three weeks with four horses and a waggon, and so we were completely cut off. But what used to come through was ' armoured car, and it used to fetch bread from Hull, but it cut across the fields, and these lads that were driving, a lot of 'em had been in ' army, it was egg and milk to them.

As the cold and snow continued, and the buses could not get through because of the snowdrifts, the people who usually took the bus to work had to walk. For those who lived on the coast at Aldborough, the route to the bus at Bilton took them past Carr Farm.

They used to be passing when we were doing horses in a morning; that was between five and six. There was men and girls mainly, and they'd come about three miles when they got to us and then they'd got to walk to Bilton, and at night they were walking back when we were doing horses again at six, and we used to have a big heap of snowballs ready and snowball 'em all! Rotten buggers we were! And they used to have a go back at us again; it was all done in good fun.

The cold weather persisted for weeks, so the only work it was possible to do was providing food and water for the livestock.

That's all we did, lead straw into bullocks and thaw pipes out. There was one tap, and one tank, and we had to keep thawing this because the ponds got frozen up. They got frozen up that much 'til it was hell of a job breaking the ice, 'cause the rule was in that place that no horse went back into stable without going into pond to get its legs washed. Well, as soon as it started getting frozen and frozen and frozen, they started to let 'em into the yard to drink with the bullocks. Usually when you get forty or fifty bullocks in a yard when they keep going to drink, they keep the tap running, but it beat the bullocks. Oh, we had hell of a job with lumps of wood, banging and messing about.

Then took two labourers who hadn't got a snow job, they had a big hedge to fell. It was a thorn hedge but it had grown into nearly thorn trees, and there was a big dyke, and when they cut 'em down some on 'em would go straight through the ice, and after they'd been going for a day or two, when we'd finished strawing, he says, 'Get a set of *draughts* and put your 'osses in a waggon and take them draughts and go down so and so.' And when we got there we put these draughts on and a chain and hung to them, and he says, 'Right,' and he rolled his sleeves up. There was ice and snow everywhere, but just where this hedge was, there was about six yards with no snow, and he just sticks his hands with this chain in this water. 'All you have to do,' he says, 'is drive them horses.' I had to back 'em, he'd hang 'em on, pull 'em out, and he did this for about two hours. So he looks at his watch and he says, 'Time we was loadening up,' and puts his jacket back on. Never said it was cold nor nothing!

And when we going for straw, we cut these straw stacks in half, and you always shoved your knife in the stack, but the handle was left out. But if ever we'd to do it, which wasn't very often, you'd pull some straw out and put a little ledge of straw on top of your knife, so when you got your knife out there was no frost or ice on it. But he just got hold of it with his hands and he'd thaw two hand holds

80

through while he went, and woe betide anybody who put gloves on. He used to say, 'I've never seen a man who could do any job with his gloves on, so take 'em off,' and even when we were going to plough you had to 'a' your gloves off, and you'd be pulling your jacket down over your hands.

By the time the cold weather finally abated, the spring work was well behind. Even when the soil had thawed enough to allow it to be worked, the snow drifts which had been blown into the shelter of the hedges took longer to melt, so one of John Caley's brothers who farmed on the Wolds at Tibthorpe had to drill some corn round a snowdrift in June.

CHAPTER 6

From Spring to Martinmas

T HE arrival of spring brought weeks of harrowing for the horses and horselads, because although the ploughed furrows had been partly broken down by the frost, the heavy soil still needed repeated passes with the harrows to get into a tilth fine enough to drill the spring corn. Most of the work was with the heavy gibb harrows with their forward-facing tines which pulled the harrows into the soil, making it hard, slow work for the horses.

> Late spring come along and we were three 'oss harrowing, there was three lots of three horses harrowing, and a horse in ' artificial drill. And when we was harrowing you could turn round [at the ends of the field] and still be in your stride, you were going that slow, and if you put one foot on your harrows and just pressed, all the horses just stopped dead, and as soon as you took your foot off and just clicked, they went again; they were loadened up to the hilt. With three horses in a set of gibb harrows you could do twelve acres, and if you put three horses in a swing chisel, you might get seven acres.

The swing chisels were similar to normal chisel harrows, except that instead of being in three sections, which can move independently and follow the contours of the land, the swing chisels were all in one piece, and being rigid, were harder to pull. Another hard job was rolling the ploughed ground. Three horses were used to pull the roller, which was also done at a slow walk so only seven acres could be got in a day. At Carr Farm, once the soil had been worked to a tilth, the corn was drilled with a tractor, the horses following with harrows. In the autumn, gibb harrows were used after the drill because the land still needed to be worked down, but in spring the seedbed was already fine enough, so a lighter set of six seed harrows pulled by three horses was used. Once the corn had emerged, the crops were rolled, but this time a pair of horses was adequate as the surface was much smoother than the uneven ploughed ground.

> Later on in the spring when everything was growing nicely, we started to roll, and when the barley had got several inches high we would sow it with small seeds. There'd be three horses on the front in a set of six light harrows, and then there'd be Foreman in the middle with a fiddle drill and then there'd be two pair of horses behind in Cambridge rollers with poles in mainly, if we hadn't smashed 'em. If we smashed 'em you just hung to the end and they were swing; there'd be no poles in, ' just pulled 'em. You could harrow sixty acres a day with three horses and six harrows, but we used to reckon twenty acres a day with a Cambridge roller. They used to come and tell you they wanted more music into them, and they'd pick some clots up and throw it under their bellies; they'd say, 'Keep 'em going like that, they're alright!' They hadn't to trot, but they were on the verge of trotting.

The small seeds sown on to the growing crop of barley were mostly white clover, mixed with a small amount of red clover, which grew slowly at the bottom of the barley crop until harvest. After the barley was cut, the clover then had more light and moisture, allowing it to grow quickly and be sufficiently well established to be grazed by sheep in the autumn. The sowing of the clover seed and the twenty acres of mangels, which was also done by the foreman, were the last tasks of spring, and shortly afterwards the horses were turned out to grass after work. This was always done gradually, to allow the horses' digestive systems to adjust to the change in diet.

> We turned the horses out to grass on May the eighth, 'cause that was when the frost was supposed to have gone. Previous to May the eighth they were turned out as soon as they come home from work, then we went and got them in at eight o'clock, just for the week, and then the week nearest the eighth of May, on the Sunday morning, we turned 'em out and they were stopping out then.

Once the horses were out at grass, the horselads had more free time at the weekends. Ron sometimes took the bus from Sproatley to see his parents in Hull, but other weekends were spent around Flinton or riding round the area on bicycles.

> In summertime, we'd bike to the seaside places, Hornsea and Withensea, they were about ten or twelve miles roughly. Everybody

was on bikes, and of course you used to pick these lasses up and we used to all go riding together. Oh, it was great fun, I suppose. And we always used to go Aldborough, go in the sea. I mean it was only three miles from Flinton, you could go in the afternoon and then come back for your tea.

And on a Sunday they all used to bike from Hull to Aldborough, and one thing we used to do, we used to make parcels and wrap it up in brown paper and put it in the middle of the road. Then we used to have a bit of string on it and sit in ' horse pasture, at back of ' hedge and you'd hear these bikes coming.

'A parcel, look!' and they'd jump up off their bikes, and just as they was going to put their hands on it, we'd pull our string and pull it. And then sometimes we used to leave the parcel full of 'oss muck. 'Ooh, parcel!' They'd grab it, like; they'd undo it and then there was a scream when this 'oss muck used to fall all over, and that was one of our usual stunts for a while 'til the novelty wore off!

One of the first jobs of summertime, once all the cattle were out in the fields, was to muck out the fold yards. Because of the copious amounts of straw used for bedding, the muck could be as deep as four or five feet. Muck leading was one of the hardest jobs for the men, so sometimes the workforce from Carr Farm was joined by the workers from Pasture House.

When we got muck out, depending where we were going to put it, you could have anything up to seven waggons if you was going into ' far fields. We used to get three load in the morning, and you might get four load in the afternoon. There was two or three men in the yard filling, and a man at the muck hill helping you to team, so he kept you right at the muck hill. Some farms wouldn't make square muck hills because the corners dried out: you had to make a round muck hill, whereas others wasn't so particular, but the hills were dead straight up. Usually the muck hills were on white clover *seeds*, and you put the muck hill in ' middle of the field depending on how big the field was. Carts were very rarely used, everything was waggons or rulleys. Sometimes they'd put a 'oss in a cart to get into a corner easier, or if they wanted to shaft a young 'oss and get him into some work. I mean, it used to take us three or four weeks to get muck out of one yard and another so if he'd got a fortnight in a cart for eight hours a day he soon got used to the job.

In addition to hilling muck, which would be spread in the autumn before ploughing the seeds for the following crop of wheat, some of the muck was spread directly from the fold yards on to the fallow land.

> If we were going to muck fallows, which they'd ploughed several times, it was just one field of clots. At one time they did it with horses; but they used to plough 'em with tractors and plough you some roads out, muck roads as we called them, right across the fields. The waggon wheels would fit into the furrows what he ploughed out, and your horses walked down the same marks, so your horse was directly in front of the wheels. Your pole chains were at full length and your helters [halter shanks] were at full length because that pole, it used to swing backwards and forwards and it used to catch their legs. They'd learn to keep out, away from the pole. And again you had somebody helping you to team, but this time you was going to spread it. You had equal spaces for the waggons and equal distances between the wheel marks, so that my waggon would spread to his waggon and each one had a road to hisself. So your horses went to the far end of the field, so that everybody was going up their own roads, and when we turned round, we put the man who was helping you to team off onto the next waggon, so he hadn't to climb about, especially if he was an older feller.

In between the muck spreading, the twenty acres of mangels needed attention. The mangels were grown in the same year of the rotation as the fallows, the main purpose of which was to reduce the number of weeds by continuous cultivation. Since the mangels were sown in the late spring, any weed seeds that germinated earlier were killed by harrowing the seedbed, but once the mangels were sown, a two-row horse hoe, locally known as a *shim*, was brought into action to remove the weeds between the rows of plants. Because the rows were so hard to see when the plants had just emerged, the horse was led by one man, with another guiding the shim, but later on when the plants were easier to see, one man could manage both the horse and the shim. The individual mangel rows also needed to be thinned, or gapped, to allow enough space for the plants to grow properly, this being done by the labourers using hoes to chop out the plants.

> We'd two-row shims, and twice over you had a leader. You shimmed 'em the first time, then they gapped 'em, so next time you

went, there were heaps in the middle where they'd been gapped, and they led you again. Then after that you was on your own, and then depending how much time there was, they'd send you to shim if there was nothing much to do. The first time over with a leader they'd maybe get about seven acres, but once we got going with these two-row shims, we could get ten acres a day with no leader, and then very often we had single *scrufflers* which would do between maybe four and five acres a day.

The scruffler was also pulled by a single horse, but only worked the ground between one pair of rows. However, it did penetrate the ground more than the shim, so would break the surface crust sometimes created by the horizontal knives of the shim. The same shim that was used on the mangels was also used to hoe the corn, except that more hoe blades were added, matching the width of the rows on the corn drill. As with the mangels, the weeds growing in the rows were cut out by men with hoes, *looking*, as it was called.

In the meantime the labourers, and us 'osslads depending on what we'd been doing, we'd been hoeing the corn. We used to take the field six rows each at a time, and you all walked in a row up and down the fields, knocking the thistles out and pulling the docks. We'd be in spring corn 'til dinnertime and then the wheat'd have got dried a bit, so we'd go in the wheat in the afternoons. Before they'd come in with their hoes we'd probably have a horse hoe, and we could do twenty acres a day with a ten-row horse shim, one man leading, one man shimming. I've seen as many as ten of us hoeing, and Foreman would be just five yards in the front, and he would keep there all day long.

Everybody used to get sick of looking, and if he thought us lads was getting too sick, he would send us scruffling for half a day and then fetch you back again, 'cause when you got sick, you wasn't always looking what you was doing. Or he would say, 'Alright, we'll go ploughing fallows,' and if they were ploughing fallows, you'd wish to hell we were back looking, cause it was a lot better job looking than it was ploughing fallows. Nobody liked ploughing fallows, and when they come to harrow these fallows, they had sets of harrows with the tines two-foot deep, a set of wood harrows, and you put a bag of straw on and you sat on these harrows, 'cause you couldn't walk, it was that rough. You'd never seen anything so rough in your life, clots all over, like horse's heads; you was stumbling, and

86

they were an' all. And that's where the hair comes in on a Shire horse's legs, so these clots couldn't cut their legs. 'Cause they did cut their legs, and you'd see 'em bleeding, and that hair cushioned them.

The work of the summer moved between different jobs, depending on the weather and how forward the crops were, so the horselads might well be sent scruffling in the morning, and go haymaking in the afternoon. In parts of the country where grassland predominated, haymaking was the most important and labour intensive part of the year, but in Holderness, haymaking was of much less significance and was fitted in between the work on the arable crops.

Well, the waggoner had been with the grass reaper, and if they'd took two pair of horses [in turn] with one reaper they could cut something like ten to twelve acres a day. Well, these farms didn't grow a lot of hay; they just grew enough for their fat bullocks, depending on how many bullocks they had. If they were cutting twenty acres of hay, we'd maybe cut it in two lots, cut ten acres and then you would start and get that ready, you turned it with a swath turner. Then when we'd turned it a time or two, we'd go with a horse rake [trip rake] and put it into heaps, as much hay as you could [put] in a cock.

Or you'd maybe go with a side delivery [rake] and turn it into a big row, 'cause [with] the side delivery you could go one way and put so many rows to the left, then come down the other side and put so many to the right, instead of having to 'oss rake it. And then the men would follow you cocking it, and you cock it so that if it rained it was reasonably safe.

The cocks in Holderness were quite large, about five or six feet high, and once they were sufficiently dry the hay was put onto waggons with one man each side of the waggon forking hay onto the load.

Then you took it to the stack and teamed it, and Foreman and two labourers would be on the stack. If we had ' elevator we only had to team into ' elevator, and you did that 'til the field was cleared. In the meantime they'd probably cut the other one, and by ' time you'd finished that one, the other one was almost ready for getting.

In arable areas like Holderness, harvest was by far the busiest time of the year. Up until the war, many Irish labourers, mostly from the north-west of Ireland, came over to work during harvest, the same men returning to the same farm year after year, bringing their sons who would then continue the tradition. They were paid an agreed amount for the month, working from seven until seven, and were all fed by the foreman's wife. However, the war put an end to this annual migration, so much of the work in harvest was then done by prisoners of war, first the Italians and then the Germans; but Ron's first harvest at Flinton in 1947 was the last year that there were many Germans still working on British farms.

Then harvest would come, middle of July, and we would start among peas. Now we always grew a hundred acre of peas, and you had to cut them the way they were laid. So very often you was only cutting down one side, what we call fetching 'em. And depending on how long the field was, you had to have a man every so many yards to move the peas, putting 'em off in nice little forkfuls, and then turn the pea heads back so that when the grass reaper come down again it was cutting ' full width of its knife.

Six weeks after the peas flower is when they're going to be ready for cutting, usually about ' second week in July, and sometimes they'd have two or three lots going. One lot might be horses and the other might be a tractor, but it was always the same principle, every time you cut 'em you had to move 'em. Then when you cut 'em, after they'd been dried a time or two, they all had to be turned by hand; ' fast as you could walk you could turn 'em. By this time we had swarms of German prisoners, we'd maybe a dozen, twenty German prisoners of war, and at one time they used to reckon some-thing like two acres a man an hour.

Every time it rained you turned the peas, and there was one year we'd turned 'em that many times for the weather, we just got the seed back. That was eighteen stone of peas for seed, after we'd turned 'em and turned 'em, and thrashed 'em, we just got what we'd sown. And the fields were full, you could see all the peas, and we used to take the pigs on then, and these little lads used to tent 'em, one sit at this side and one at other and keep 'em in, and then take 'em home. That's if the peas were adjoining the farm yard, they used to take 'em home at night and put 'em back in ' fold yard.

When they got started among peas, we'd also start among taties. We started *green topping* some in July, that's while they're all nice and green, and I would have three horses and a tatie spinner and I'd be tatying when the others were among peas. I'd have a gang of German prisoners and there would be another lad with the scales. We had the scales on a sledge, and he would go round as they were picking them and weigh the taties in hundred weights, then drop 'em off in fives and the lorries would then come in and pick 'em up. Then we'd have to loaden 'em with a *hicking stick*, straight onto the lorry. In those days they used to come for five ton, and providing it was reasonable, the lorries would come in the field, and if it got a bit wet I would very often have three horses hung to the front of the lorry, just helping it along.

Lifting the potatoes, harvesting the peas and starting to cut the wheat were all happening at the same time, but as soon as the peas were dry enough, they were loaded onto waggons and led to the stackyard.

When the pods were brown and the peas were bullet hard then we'd come in with the waggons, and you'd to put all these peas back and make roads so that the waggons could come between 'em. There was all the rows of forkfuls of peas, so the forkers could just put his fork in one lot and put 'em straight on to the waggons.

'Course, if you'd hundred acres of peas, you'd at least ten days threshing, and they'd be very often put in what they call *coupings*, each stack was an individual, but they were all joined up. Then when you'd got these peas stacked you wasn't long before you *happed* 'em down with wheat straw and put a coconut net over 'em and tied your bricks on. And then Foreman at Pasture House, whose name was Charlie Buck, would thatch with wheat straw from the floor to the eaves, so the peas wasn't getting bleached, 'cause when they got bleached on the outside they were spoiling the sample. Now the other farms what didn't thatch, they used to put some sheets down and thrash the outsides with sticks and then bag 'em up, and bring the bags out and put 'em through the threshing machine on threshing days. It wasn't everybody who could thatch.*

Well, by now some of the corn would be ready. So the binders would be going, and [on] most of the farms we'd now got tractors in 'em, because if anything was going to tire horses it was a binder. Them what had horses in 'em, they used to change 'em at regular

---

* Harry, Charlie Buck's son cannot remember his father doing this, but perhaps Ron was remembering something Charlie Buck had told him, rather than what Ron had actually seen.

intervals, and of course you never got started while the dew had gone, and by about seven o'clock the dew was falling again and they'd cut about ten acres a day, if it was good going.

When cutting with a binder, the corn is cut before it is fully ripe, so that the grain is not shed out of the ear as it passes through the binder mechanism. After the binder had thrown the sheaves out onto the ground, the sheaves were all picked up in pairs, one under each arm, and stood up in the stooks, so the grain was kept off the ground and could dry out. The corn was always stooked with the axis of the stook running north to south, so that both sides would receive the benefit of the sunshine through the course of the day.

> You always worked in pairs when you stooked. With a six-foot binder, you took six rows, the two of you, and when it was ' eight-foot binder, we took four rows for the two of us.

On the farms where Ron worked the stooks of wheat consisted of no less than six pairs of sheaves, though in other parts of the country with different weather patterns, the number of sheaves in a stook varied. The oats and the barley, however, were stooked in tens, while two pairs made up the stook for beans, the first pair being stood up as normal and the next pair put at right angles to the first, so there was one sheaf in each quadrant of the compass.

Until 1947 all the corn was cut with a binder and stooked, but the following year John Caley bought the first combine.

> When the combines first come, it came in packing cases from Canada, and they came from ' Burton Pidsea Plough Company to assemble it. Them was the Massey-Harris twenty-ones, twelve foot, and they had canvasses, two feeding it and one taking it back up, and they had *sails* on. And then they got off sails, and they got onto pick-up reels. I used to have to go and help him to grease up, and I think there was something like a hundred grease nipples on that pick-up reel.

Initially the combines cut the barley, which was uncomfortable to stook because of the awns, and more importantly, it was difficult to stack because it was so short and slippery. Also all the corn which had been blown down by the wind was combined because it was hard to cut with a binder, and hard to stook because the sheaves were very irregular with bent and intertwining straw.

> Round here there was acres, as far as the eye could see, it was all laid
> as flat as could be, was all the corn; flat as your hat, wherever you
> looked.

The introduction of combines solved some problems, and saved
some labour during harvest, but it also created a new problem.
Before farms had driers to dry the grain, often the bags of corn from
the combines were too damp to store safely, so after they had been
picked up from the fields where the combines had dropped them, the
corn had to be emptied out of the sacks onto the granary floor where
the air could get to it. This was a problem not encountered with corn
that had been in a stack, as the air between the sheaves allowed the
grain to dry slowly throughout the autumn and winter.

In a normal year the sheaves of corn would already be fairly dry
when stacked, having been in the stook for two or three weeks. The
tradition was that the stooks of oats, which are most susceptible to
going mouldy, should hear the church bells on at least three Sundays,
and so have plenty of time to mature and dry before being stacked.

The leading of the sheaves to the stackyard was traditionally done
with a pole waggon, but in 1947 John Caley bought the first
four-wheeled trailer, made by Tye in York. Although made for use
with a tractor, at Carr Farm it was pulled by a pair of horses. Unlike
the Yorkshire waggons, these trailers were fitted with harvest ladders,
or *gormers*, which made loading the sheaves easier.

> Then this Tye trailer come along, it had gormers with it, and then
> another two came and one had got gormers on, but the other one
> hadn't. And when they came with gormers we said, 'What do you
> want gormers for? We don't need 'em,' because a Yorkshire waggon
> has no gormers, and no rails nor nothing, and it was a lot harder to
> load them than it was them Tye trailers. But after about a fortnight
> we wanted to know why the other one hadn't gormers, 'cause they
> all want gormers!
>
> So then we had three trailers, instead of as it had been done, very
> often they had seven waggons to two stacks. So there was one
> at each stack teaming, two loadening, and three somewhere in
> between. But now we got three trailers and there was two forkers
> and two loadeners and a pair of horses in the trailers. We just used to
> hang to the drawbar and pull on, and the drawbar stuck in the
> ground, that was how it stopped you, and I remember on one occa-
> sion pulling the pin out and moving off, and the very trailer set off

with a full load on. It went down this incline and I thought any moment now it's going, but it didn't, the drawbar stuck in the ground and it didn't turn over.

Once the loads of corn arrived at the stackyard, there was a routine to be followed.

The first thing you did when you came into ' stackyard, there'd be one man just teaming, and any sheaves what'd fallen off, the first thing you did was go and stick your fork in and throw these sheaves up to him. Then you went to your own load and took your waggon ropes off. When he pulled out, you pulled up. Now you could have maybe three forks stood against your waggon onto the sheaves. There'd be the fork you went up with to start with, and when they were starting the stack you would use this fork, 'cause you were throwing them downhill, and this fork did you for a while. Then you'd got down a bit, and you just put that one down and you got your next one, and forked, and then you had your Long Tom, which was a six-foot shaft.

Once the stack was as high as the forker could reach with the Long Tom, the stacker would then leave a pick hole for the picker to stand in. On some farms, however, instead of leaving a pick hole, the picker was provided with a *monkey*, a platform about three or four foot square which rested on one side of the stack.

At Pasture House they always had a monkey, and they're a platform with two legs; the platform protrudes out, and you put a rope on this monkey and you put it through the stack and tie a four-stone weight on it at the back, and they stack over the rope.

When the stacks got to this height, the wind often made stacking difficult. Not only was it harder to fork the sheaves up, but they would also blow off the elevator, and the wind would lift the stacked sheaves on the windward side of the stack. In sheltered areas the wind was less of a problem, but the open landscape and strong winds around Holderness meant that stacking was often a difficult job.

There isn't a lot of trees. There's just these little clumps around the farms as wind breaks, 'cause it always blows round here. In harvest, we used to have to have sheep hurdles tied on the elevators to keep the sheaves going up, and on the stacks we had sets of harrows, and when you were going round a corner somebody used to have to lift the harrows up while you were going round. Oh, it'd be nothing to be stacking in a top coat!

In Ron's time, the stacks usually had a rectangular base, sometimes with rounded ends, but in earlier times completely round stacks were built, known as *pikes*. These had to be thatched to keep out the rain, but by the 1940s most farmers covered their stacks with loose straw rather than thatching them in the traditional manner, so pikes went out of favour. Nonetheless, there was still variety in the way different people stacked. Stacking the sheaves in a stack is a skilled job, as every sheaf needs to be laid sloping outwards to shed any rain, and each row of sheaves is partly covered by the next row, which stops the sheaves slipping. Stacking was therefore often done by the foreman, and each foreman had his own style of stacking.

The most common style of stack in Holderness was rectangular with round ends. With any corn stack, each outside row of sheaves was put very slightly further out than the last one, so that at the eaves any water from the roof would fall clear from the stack. At the eaves the courses were then drawn in to form the roof shape, and in the case of the round-ended stack, the ends were brought in as well. At Pasture House, however, Charlie Buck made house-ended stacks, which had a rectangular base, and the ends went straight up to the ridge. Although this was harder to stack, the roof held more corn than a round-ended stack, and it also required less straw to cover it. George Gibson at Flinton, meanwhile, made stacks which were rectangular, but the corners were rounded slightly. The ends of these stacks, like the house-ended stacks, also went straight up to the ridge, but just as the corners were rounded at the base, so too the ends of the roof curved in a little bit.

Covering the house-ended stacks with straw only required straw to be put on the two sides; on George Gibson's stacks the straw went round the ends a little as well; but the round-ended stacks needed the most straw as the ends also needed to be completely covered.

> Years ago they made pikes, which was a complete round stack. But when they started to cover 'em up with straw instead of thatching, they couldn't cover 'em up very well, and whereas they tell me it used to take 'em all day to thatch a stack, and there was also somebody getting the straw ready for 'em, we could cover four stacks down a day. We used to get a waggon load of straw and they used to fork it up to us, or when we got elevators, they used to put it up '

elevator, and they could be happed down quite well, and they were completely waterproof.

When happing down a stack with loose straw, the straw at the eaves was put on thinly, the layers getting progressively thicker as it went up the stack. The man on the stack walked backwards, working sideways and upwards, using his fork to place the straw, and over-lapping it in both directions until about three-quarters of the way up. Once he had worked his way right round the stack, he then capped off the top to a thickness of two feet, making sure he did not stand on any of the straw.

They was all happed down just after harvest, and it was very nice if you could get on 'em when there was some dew on there. But as the day got on and the sun got out, you'd get it all nicely on and all of a sudden she would go, and it'd all slip off, and if you was anywhere near it you went with it. And those stacks, when we used to finish them, they used to put a coconut band net over. They used to wrap these nets up in a special way and inside these nets were four waggon ropes, and it was all wrapped up as a parcel, and then they used to stick their fork in and carry it up a ladder, and fork it up. Then if there'd been two on you when you'd been doing it, one getting the straw to you, and you were happing it down, as soon as you'd finished, the other man got off, and while you were doing this they were tying string on bricks all ready, and you undid your net and you threw your waggon ropes down at every corner and you got your legs through a mesh in the net. They were quite a big mesh, and then if there was four of you handy, they'd get a waggon rope each and work it and pull it, keep levelling it and you'd go with your fork, and then when they got it right they would tie a brick on each corner. Then we'd go and get another load of straw and leave somebody to tie the bricks on that stack, about every two foot, and they were secure for the rest of winter.

By the time harvest finished, the main crop of potatoes was ready to be lifted, the skins on the tubers having set, so they could be safely lifted and stored for the winter. The potatoes were all spun out of the ground with a potato spinner, then picked into baskets and put in carts to be taken to the potato pie, in a similar manner to the mangels, which were the last crop of the year to be harvested.

Lifting the mangels started a week or two before Martinmas, with the aim of completing the job before the lads left. The mangels were all pulled by hand, two men working together and taking four rows each. Each man pulled a mangel up by its leaves with one hand and cut the leaves off with a *slash knife*, letting the mangels drop into a row, so there was enough room for a horse and cart to go down between the rows.

> We didn't used to start wuzzeling while mid November. If we were going to pie in the field we only needed about two or three carts on, one man at ' pie and two or three men pulling. When we went in a morning, everybody would pull, and we would go in at back of 'em and we would start to fill [the carts], and then they would come and help us to get us loadened up. Then they would go back to pull, and maybe two men'd be constant pulling. We used to drop 'em four rows into one, and they just left the room between for a cart, and then we used to stick 'em with our forks and shove 'em in. I've a scar or two on me hand when our hands have been that frozen, 'cause very often it would be freezing. It was only just us lads that seemed to bother. Them labourers didn't seem to feel the cold, they just kept pulling, and you had a bag round you to stop your trousers getting wet.

Harvesting mangels was one of the few jobs in Holderness where carts were used in preference to waggons, as the carts could be tipped up at the pie. On lighter land one horse might be sufficient to pull the cart, but in Holderness two horses – one between the shafts, and another, known as a first horse, in front – were always necessary when wuzzeling. Even so, getting the load to the pie was difficult, even with the newer carts which had pneumatic tyres, as the carts soon sank down to their axles in the soft soil.

> Some [carts] were on rubber tyres, some were on irons. You had to have a different set of ruts for the iron'ns, and a different set for the rubbers, and when we've been taking 'em home, I've seen seven sets of ruts. We've got that deep we couldn't use 'em, and they had to keep making fresh ruts, and you had your raincoat on back to front and when you got anywhere near where you all converged into one, it was just like a sea of liquid mud, axle deep. The biggest job was when you got to the pie, backing 'em, because it was no easy job. When you brought 'em up to the pie, you cut right across the front of the pie, turned 'em short round and you took the first 'oss right

95

round 'til it was facing the back of the cart. So you'd the first horse in your left hand, and your shaft horse in your right hand and you told it to back and it to pull. So when it pulled, you backed, and you got back to the pie.

Once the mangels were all in the pie, it was covered with a layer of straw about two-feet thick. Then when the straw was all on, a single horse was used to plough a furrow up against the bottom of the straw to keep it in place, and then more furrows were ploughed round the pie to loosen the soil which was then shovelled up on top of the straw to a depth of eight inches. Rather than throwing the soil a long way in towards the pie, the men also dug downwards, but the further down they went, the further up the soil needed to be thrown. This meant that only the bottom three-quarters of the pie was soiled down, the top being capped with dyke reapings.

The mangels were the last crop of the year to be harvested, and for the horselads it was often the last job they would ever do on that farm, and the last time they would drive those horses before leaving at Martinmas.

CHAPTER 7

Checks, False Lines and Pole Waggons

WHEN the use of horses was widespread, many different styles of harness were developed in different regions of the country. Part of the variation was due to the stylistic traditions of harness-making in each area and the preferences of the harness-makers and their customers, but there were also functional differences dictated by the requirements of different vehicles and the working practices of the different areas.

In East Yorkshire, for example, the harness used when going to plough, or when a horse was put in a cart, was similar to that used in the rest of the country, whereas the use of pole waggons required a completely different type of harness which was not found elsewhere.

Local methods of working, specifically how horses were hitched together, also influenced the harness used. In Yorkshire, horses always went to work with a hemp halter underneath the bridle. The reason was that except when a horse was worked on its own, the rope from the halter – the halter shank – was used as part of the steering mechanism to tie each horse to its neighbour. So when working a pair of horses abreast, each horse's halter shank was tied to the other horse's hame ring, or to an unused trace link hanging from the *hame hook*. These crossed halter shanks, which were six feet long, were easily adjusted by tying them at a different length. They not only regulated how far apart the horses worked, but also prevented a keen horse from getting in front of its mate, without any pressure being put on the horse's mouth. (See Figure 1.) In North Yorkshire it was also customary to have a light rope joining the inside bit rings of each horse, so the pull from the driver's hands, via the *cords*, continued through the bit of one horse, along this coupling band to the other horse's mouth. But in the East Riding no coupling band was used, so when a pair was turned to the right, for instance, the offside horse

Figure I   The usual method of driving a pair of horses in the East Riding: each horse has a side string to its outside bit rings and is tied back to the other horse's hame with its halter shank.

reacted to the pull on the cord from the driver, while the nearside horse had only the halter shank to pull it round.

The use of the cords, certainly by some horsemen in Holderness, was also different from elsewhere. Whereas some simply tied the end of the cord to the bit ring, which is usual practice throughout the country; others threaded the cord through that bit ring, passed it under the horse's jaw and tied it to the other bit ring. So on a near-side horse of a pair, the cord went through the nearside bit ring and was tied to the offside bit ring. Other people in the area did it slightly differently; after passing the cord through the nearside bit ring, it went through the offside bit ring before coming back under the jaw again and being tied to the nearside ring. This meant that the pressure of the cords on the horse's mouth was somewhat different from the practice in other regions, but it was what the horses were used to. A variation of this last method was also used in the show rings.

> When the fashion in the show rings in the country shows in the East Riding was for the pairs to be driven in plough gears, just in the harness they went to plough with, they used to have white strings and they put them through as if we were going to plough, but pull them

straight through, and they went halfway [back] down the horses' sides and then they were finished off with a *bowline*. So each horse had a double length of white cord showing, just for appearance sake.

Even when a horse was used on its own, the nearside cord went under the jaw, and was tied as described, while the other cord went directly to the offside bit ring.

The hemp cords used for everyday work, known as *side strings* in Holderness, usually passed through a ring between the bit and the driver's hands, in order to keep them off the ground and away from the horses' legs. In some parts of the country the back bands had a ring stitched onto them for this purpose, but otherwise the cord went through the hame ring, especially when the horse was being driven from a vehicle, so the line between the bit and the driver was fairly straight. While walking, especially with a plough, when the horse-man was walking in the furrow a few inches below the soil level, many horsemen preferred to thread the cord through a dropped trace link at the hame hook, as this lowered the line of the cord, so a more direct contact was made with the horse's mouth. In Holderness, by contrast, many horselads made a loop of leather which hung from the hame ring, and threaded the side string through this loop. Besides lowering the cord, it also lessened the wear on the hemp as the line went back and forth during work. When working the same pair of horses day after day, rather than removing the side strings and coiling them up at the end of each day, the horselads often left them on the collar, so they would be there the following day.

When three horses were worked abreast, the horses were fastened together in a similar manner as a pair. The cords went to the outside horses, and the halter shanks were crossed in the same manner as a pair, but for the middle horse to be tied back to both the other horses, another halter shank was needed on the middle horse. In practice this meant putting a second halter on, because although the halter and shank were made separately, after a little use they became so tightly pulled together that it was impossible to take the shank off. When four horses were yoked abreast, the two horses in the middle both needed another halter for this set up to work.

Holderness also had another completely different manner of driving; instead of using a pair of cords, drivers used a single line to control a horse. This was called a *check*, or a *check rein*, and a horse that was trained to work in this manner was called a *check horse*. The use of the check rein was confined to Holderness and the Fens, though the detail of the Fenland driving differs somewhat from the Holderness style. Driving with a single line was also common in the low countries of Holland and Belgium, and it was probably from there that this method of driving came, when Dutch engineers were brought in to drain the land. This style of driving was quite distinct and never spread very far from these particular areas.

> Round here they'd a check and a *false line*; if you see any old photos of people round here, all these horses have a check. Now the further you get to where it's lighter, the less they have checks. I never heard much about 'em driving them on the Wolds in these checks, but I could be wrong. But certainly the further you got down here, as the land got stronger they used checks more.

The check rein consisted of a length of leather, the front of which split into two, one piece going to each side of the horse's head, the nearside piece being shorter than the offside. These both terminated in a length of chain with a clip on the end. Although not necessarily the case on other farms, in Ron's experience the nearside chain was put through the nearside bit ring, passed under the jaw and clipped into the offside bit ring, while the offside chain went through the offside bit ring and was clipped back into itself. The most common type of bit used on check horses was a curb bit and chain, though some wore plain 'bucket handle' snaffle bits. The rear end of the check rein went between the tops of the hames, where it was then formed into a loop, onto which was tied the cord that went to the driver's hand. The end of the cord held by the lad was also formed into a loop, and was significantly thicker and stiffer than the rest of the cord, extra material being added to this end when the rope was made. On some farms where the farmer had not bought these tapered plough cords, the lads simply spliced on a length of old halter shank themselves to make this loop.

In use, the check rein was pulled for the horse to turn to the left, and given two or three short jerks for the horse to turn to the right. Much of the control over the horse, however, was by the voice, the severity of tone indicating how much the horse should turn.

> You pull to go arve, and you only just had to shak yer rein and they go gee back; and you talked to 'em a good lot, and the more severe your voice was, the more they knew they'd to keep coming.

When a check horse was used in a team of horses, it was the key to controlling the whole team. In Holderness it was usual for the check horse to be on the nearside, with the offside horse tied by its halter shank to the check horse's hame ring, and a false line to the check horse's trace. A false line was similar in construction to the check; the front was divided into two parts, with a length of chain at each end, which clipped onto the appropriate side of the horse's bit, the offside chain being five inches long, and the nearside six inches. The piece from the nearside passed under the horse's throat, where it joined the offside piece, passing over the offside of the collar, and over the back of the horse, before being clipped onto the check horse's trace just in front of the swivel. So when the check horse slowed down or stopped, pressure would be applied to the false line, and thus to the other horse's bit. When turning, the check horse brought the other horse round to the left by the pull on its halter shank, and when moving to the right would physically push the offside horse round, if it had not already responded to a vocal command.

When ploughing with a check horse, Ron's experience was of only having one line going to the check horse, with the other horse controlled by the false line, but when harrowing with three abreast, two reins were used. The check horse was on the nearside, the middle horse's halter shank was tied to the check horse, but it also had a side string to its offside, to help bring the team round to the right. The offside horse then was tied to the middle horse with its halter shank and had a false line on it. (See Figure 2.)

In practice different people did different things, so although Ron had the check horse as the nearside horse when harrowing with three horses, one of the photographs of Ted Simpson not only shows the check horse in the middle of a three-horse team, but also shows that there is a side string on the nearside horse. Ron also never tied the halter shank of a check horse to its neighbour, but there were some who did.

> I was talking to a feller and when he was driving check 'osses, when they were rolling they always tied the check 'oss' head in. For why I don't know. He didn't know, but that's how they did. But if they'd been a good check 'oss he wouldn't have liked it.

Even when not using a check horse, the false line was used by some horsemen, for instance when driving four abreast. In this case the middle two horses were driven as if they were a pair with side

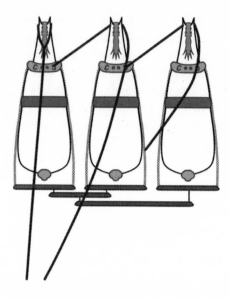

Figure 2 One method of driving three horses when harrowing. The check horse is on the nearside, the middle horse is tied to the check horse and has a side string to its offside, while the offside horse is controlled only by its halter shank and a false line.

strings, while the outside horses were controlled by being tied in with their halter shanks, with a false line clipped into the trace of the nearest horse.

Under working conditions, driving horses successfully requires some adjustment to the harness and driving methods so that all the horses work comfortably. Some horses, for example, were slow to learn to respond from the halter shank when turning, so often a loop of string from the bit ring was put round the halter shank, so the movement of its neighbour would also effect the bit of the unresponsive horse, but once the horse had become used to what was wanted, this piece of string was dispensed with. Sometimes with a young horse working in a pair, an extra cord was put on the older horse, so it would have a cord from each side of its bit, while the young horse only had one to its outside. This allowed the older horse to be steadied without putting any more pressure on the young horse's bit, and also helped when turning.

Although the use of the check rein was very common in Holderness, not all horses were check horses. Some farms did not have check horses, and of those that did, perhaps only one in three was a check horse, usually the most responsive horses which would readily respond to verbal commands. The other horses could be driven with a pair of reins, of course, but even on farms where the check was preferred, having a third of the horses in a check did allow much of the work to be done in this manner, as in a team of three horses, which was common for harrowing as well as ploughing, only one of those three would need to be in a check.

Since the use of check horses was not universal, learning to drive a check horse was a skill not all horselads had acquired when they moved to a different farm. So Ron's first experience of driving Cobby, which was a check horse, during the freeze of 1947, would not be uncommon.

> The first time they sent me up with three 'osses, Cobby was on front as a first 'oss. You rode the nearside horse, of course, but the horse we had on the front was the <u>real</u> nearside horses, they was the best horses, so you put them out front as a unicorn. I was on my own, I'd never driven with a check in me life and I got in this gateway, and for some reason we got stopped, and he come short round on me, and he ended up looking through the gateway the opposite way to what we were coming. I could touch his ears and I was sat on my nearside horse, and would he do as he was he was told? Hell, we had a pantomime, and I kept looking round to see if anyone was looking, 'cause that was the last thing you wanted, people to see you'd got in a predicament.
>
> Course, what I'd done was I started him off and not stopped him, and the horse he was, it wasn't nowt to him to come up ' side of ' other one.

One instance when a horseman always needs to be careful not to turn a team of horses too sharply is when turning at the headland with a set of harrows, but with a check horse it could be more difficult to control the turn.

> Most of them horses, if you started to turn 'em on the headland, if you didn't do anything, they'd come right round and come back on their harrows. I mean, we only had five yards to turn 'em in. When you were young, if you didn't stop 'em, the harrows would stand up on their ends, and I've seen 'em go straight back onto 'em, on their heads; get all the three harrows up in the air and all go down on 'em.

Despite this sort of problem, for some horsemen the advantages of using a check rein outweighed the disadvantages, particularly when the horseman had to be able to adjust an implement while on the move, so those like Ron who liked using check reins were keen to promote their advantages.

> Where these checks came into their own was when you was doing row-crop work, 'cause on a lot of occasions when we were among taties, we had two horses in tandem, and if we had two check horses, one in front of the other, we had just the one line on each horse, whereas if you hadn't the checks you'd have to have two reins off each horse in your hands. Very often if you hadn't two check horses, you had a check on your fost horse and a pair of strings on your second horse, because when you were two in line among root crops, when you got to ' end, your 'oss on the front had to turn while your other horse got out, 'cause your headlands were only five yards. So you imagine your two horses coming out, you had to pull and check to make him stand, while the other horse pulled out, and then you turned and come back into your row again. Once you got going, you only had to talk to 'em, and when we were scruffling or shimming root crops you just had your check, and it was considered very good.

However, there were some farmers and foremen, including Charlie Buck, who would not have a check rein on the place, considering them to be inherently unsafe. This was because unlike normal driving when pulling on both reins stops the horses, pulling on a check just caused the horse to swing round to the left, so stopping a check horse relied heavily on voice commands, which might not be sufficient to stop them in an emergency. When relying on voice commands to control a horse, it was particularly important to remain calm, so when ploughing the first time with Cobby, Ron was given some advice by George Gibson.

> I remember the first time I took 'em to plough, and Foreman says, 'You won't swear at that horse, will yer?'
> I says, 'No.'
> He says, 'No! He'll show you something if you do,' and [I was] ploughing away and when it come to finishing the furrows off, and he was a very fast horse, there was nowt much to pull and he started!
> 'Go steady, go steady; steady yer bugger, steady!' And by, his head would go up, and he stood nearly eighteen hands which was very big in those days, and he used to set off to trot. And when

A side string coiled up on the hame ring. After the string had been untied from the bit, the length of cord behind the collar was first coiled up until a yard away from the hame ring. Then the front portion which went to the bit was laid alongside the yard left between coil and hame, and both wrapped around the coils. When a few inches from the hame ring, a loop of both cords was pushed through the top of the coil, and finally the handfast (the permanent loop at the back of the cord) was passed through these loops to make it secure. This hame also has the extra hook below the hame hook for the pole chain on a waggon.

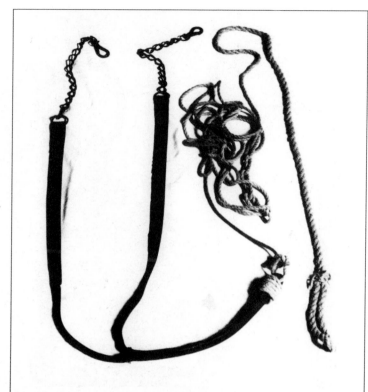

Check rein.

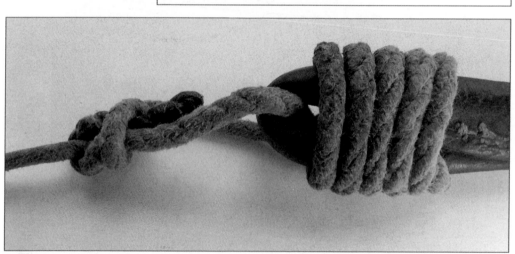

Detail of cord tied to the check rein. Instead of merely tying the cord to the leather loop of the check, which would pull on the stitches in the leather, the cord was first passed through the leather loop, brought forward, then wound a few times around the leather before passing in the opposite direction through the leather loop and being finished with a bowline.

The method of tying cord so that it does not touch the horse's jaw.

Unknown horselad, possibly one of Ted Simpson's family, ploughing with a pair. The hemp halters under the bridle are clearly visible, as is the bowline used to tie the halter shank to the hame ring. The side string on the near horse passes through a loop, possibly not of leather, to keep it low down. The use of a single horse brass was perhaps because of the presence of the photographer, though some foremen, like George Gibson, required each horse to wear a head brass every day.

Probably Ted Simpson harrowing with three abreast, circa 1900. The four reins underneath the nearside horse's head, three of which show a length of chain, show this horse is wearing a check rein as well as a hame rein. The black line above its back is the rear portion of the check rein. The middle horse is tied to the check horse with its halter shank, and the offside horse is tied to the middle horse in the same manner. The offside horse also has a side string back to the horseman.

Ted Simpson, ploughing with two horses and a wooden plough, the nearside horse with a check rein, and the furrow horse with a side string, which passes through a leather loop from the hame ring. The horses are not wearing back bands to reduce the area which will sweat.

Ted Simpson's brother, probably, ploughing with three horses, again without back bands, and with an iron plough. In this picture, the middle horse is the check horse, and the nearside horse has a side string, which looks as though it is not tied to the nearside ring, so must pass under the horse's jaw. Both outside horses are tied back to the check horse with the halter shanks, and it appears that the check horse may also be tied back to the furrow horse. The use of wooden swingletrees, master trees, and the three horse-baulks, was usual in Holderness.

Ron at West Newton with a pair of horses with heavy waggon traces.

The horse pasture at West Newton.

Cattle in the fold yard after some of the muck had been removed during the winter.

Horses in cart gears eating from tumbrils on Crofton Hill when lifting potatoes in the neighbouring field. Yorkshire cart saddles do not have meeter straps, so the horses had to get used to the collar sliding down their necks when eating.

Ron with Tim Caley, aged about 10, in the stackyard at West Newton, about 1956.

Ted Simpson riding postillion with a waggon loaded with oats. He is holding the check rein of his nearside horse. The outside trace hook is clearly facing down. Photo circa 1905.

Unknown Holderness waggoner, taking a load of grain off the farm. Since he was on the road, the horseman had leather reins instead of side strings, and horse brasses on the horses' heads. The number of sacks on the waggon suggest that the contents were peas or beans. (See Appendix 3.)

Norman Caley and his waggon loaded with 100 stooks.

Charlie Buck with one of the last horses broken in at Pasture House.

Ron at West Newton with a young horse that he was breaking in, in front of a trailer load of straw. The foreman's house is in the background.

The shepherding outfit with Graham Wilcox driving and his brother Stanton sitting in the body of the cart.

Ron feeding sheep with a Nuffield tractor.

Old Lodge, Burton Constable, where Ron and Nancy lived when they were first married.

you've a check on, you had to check and pull and check and pull; if you pulled they come straight round, you can't stop 'em, that was why this other foreman didn't like em, 'cause if they started, you can't stop 'em!

Anyway, he suddenly appeared and I was going nicely then. He says, 'You've been swearing at that 'oss.'

'I haven't.'

'You have. When I was coming across there you was nigh on trotting.'

'Aye, I couldn't get him steadied up.'

'No,' he says, 'you'd been swearing at him. I told you not to swear at him. That horse isn't used to being sworn at.' And it was quite right.

Although this incident ended happily enough, it was not always the case. One such incident was related to Ron by George Caley, one of John Caley's brothers, who by this time was farming at Little Humber near Paull.

There were three on 'em ploughing at Bush Farm, and they all had checks and false lines on. And ' Monday morning, a duck got up, and one set off, and by that, they were all gone! And all you can do is pull, and as soon as you pull you come left, so you're pulling and checking. I mean if you keep pulling, if you're up the dyke side, you'd pull him straight into the dyke. But it didn't stop nobody using them, and when they were going to Hull, if the policemen spotted them at ' tram ends he used to make 'em take the checks off and put side lines on.

Although the days of taking corn into the mills in Hull with horses and a waggon had past, the waggons were still in common use on the farms. Although the pole waggons are often referred to as a Wolds waggon, they were also used in surrounding areas, to some extent in the Vale of York and North Lincolnshire, and throughout Holderness.

The harness for a pole waggon consisted of collar and hames, a bridle, usually with blinkers, a back band, belly band and the pole britchin. The traces were heavier than normal traces and when hooked onto a *swingletree*, the outside hooks faced downwards, and the inside hooks faced upwards.

That was so they didn't get fastened in your pole. A lot of 'em had little short chains in to keep your swingletrees cocked up on top of that pole. See, waggon traces were different to all the other traces, a lot heavier. These waggon traces, we never used to go to plough with, or harrow, or anything. They were twice as heavy as an ordinary chain and where your hook was, they had a little nobble underneath, and so when you stopped it was less likely to drop off.

The pole britchin was like a normal cart britchin except that it had longer *meeter straps* to buckle onto the collar, and long chains on each side which hooked onto the hame hooks after the traces had been hooked on. The rest of the backing mechanism consisted of the pole chains, which went from the end of the pole to each horse, hooking into the Yorkshire-pattern hame hook itself, or on some hames there was an extra hook just underneath the hame hook especially for the pole chain.

The pole chains were three feet long, so the horses could walk wide apart, and so when traversing bumpy ground, the pole would not hit their legs as it swung from side to side. The long length of chain also allowed the pole to go down nearly to the ground, such as when the waggon was going down into a dip, and the horses had started to go up the other side, thus preventing the pole from breaking. Because the pole on a Yorkshire waggon fits into a socket in the

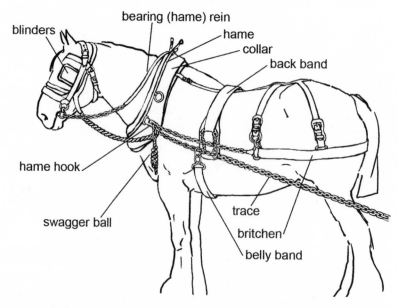

Horse in pole harness.

shears, limiting its movement up and down, when a waggon was being slowed down, the line of the pole chains and the britchin became roughly horizontal, effectively slowing the waggon. When descending steep hills, however, an iron shoe, fastened to the waggon body with a chain, was put under the nearside rear wheel to prevent it turning, to increase the friction and thereby slow the vehicle, with another chain being put round the wheel in case the wheel jumped out of the shoe.

In Holderness, however, not only were britchins rarely used, but driving without a pole was also a frequent occurrence. Much of the time this was on the farm when the soil provided some resistance to the forward motion of the wheels, so the waggon would come to a halt. On a hard road, however, the waggons would continue to roll, so the horseman had to keep the horse going or allow the vehicle to hit the back of the horse.

> I was saying about pulling these vehicles swing, no pole, no shafts, you just pulled 'em with the chains, and when you stopped, the horse stopped and the vehicle hit the horse; and these horses were quite used to this. And they've sent me down the road with 'em, first you was at this side of the road and then you were at that side of road, and it was a bus route!

Although some horses, especially the older ones, were used to having vehicles bumping into them, others did not like it at all, with inevitable consequences.

> I had a big dapple grey gelding, and I would take a load of straw from Old Farm to the next farm which was only about two hundred yards on the road, and I shouldn't have took it with this horse really. I took this load of straw and he was fidgeting, and you stopped; the waggon was touching him, and he didn't like it; and on the way back the rulley was empty and he started, and he was getting edgy, and edgy, and all of a sudden he took off and he'd gone! He [went] down the road, turned in at the stackyard. There was a tractor stood there and the rulley hit the big tractor wheel and bounced off, and he [went] full hell down ' stackyard, and he went in a big circle and he turned. ' Stable door was wide open, which it always was when the horses come out, and he run in there, and the rulley of course wouldn't go through, and there was noise like a gun going off, and both the traces broke and he flew down the stable the length

of eight stalls and he hit the wall at the other end with his head and he knocked hisself out! And he laid there on the floor and it took him a few minutes to recover, and ' course we just put him in his stall and took his harness off, but that was one of the bad things that can happen running these [vehicles] swing.

Even when there was a pole in the waggon, it was still very common for no britchin to be used. Indeed Ron never used one on a waggon when he was working in Holderness. Although the land is flatter than on the Wolds so there is less need for braking, there are still 'rises' which though not long, are quite steep.

When driving a pole waggon, there were three common positions from which to drive. On the farm, the horses were often driven while walking at their side, especially when it was cold and the lad wanted to keep warm. In harvest when the lad was on the waggon loading sheaves, the horses were also driven from the ground, the forker picking up the cords when necessary, such as when driving to another row of stooks. In this case a continuous line went from one horse's bit to the back of the horses where it could be secured on the swingletree hook, and forward to the other horse's bit.

For other jobs the horseman drove while standing inside the waggon. This was the position used when leading muck, the front nearside corner of the waggon being left empty so the lad had some-where to stand. Once in the field, when driving down the muck roads in the fallows, the strings were often looped onto a large nail driven into the underside of the *shelving* expressly for this purpose. While the horses were going down the straight muck roads, this practice was not likely to cause problems, but in other situations when the horses could turn, fixing the strings to the body of the waggon was dangerous, because if they turned, they could quickly get an enormous amount of pressure on one side of their mouth, which could result in an accident.

The other position for driving the horses, which was peculiar to the areas using pole waggons, was driving while riding the nearside horse. This was done when the waggon was full, such as when bringing a load of sheaves home or taking a load of grain to the mills. Usually the horseman rode bareback, though there were some saddles made especially for the purpose, the accompanying britchins having shorter meeter straps to buckle onto the back of the saddle.

Another unusual load requiring the horseman to ride the nearside *wheeler* rather than stand in the waggon with the load, was the waste parts of the fish landed in the docks in Hull.

> When Charlie Buck was single foreman at Carr Farm, this was about 1900, they used to loaden up wi' straw the day before, then they used to put the horses in the waggons at twelve o'clock midnight and pull out of Carr Farm and set off for Hull. They used to reckon that it was about seven and a half miles, was Hull from Flinton. Now when they got nearly to Hull there was some small holdings what used to grow lots of strawberries and they used to drop this straw off, team the straw for these strawberry fields, and then they used to continue over the river Hull, right up to the St Andrew's dock, to ' fish docks, for fish heads and fish guts, loaden all up, put it all on with forks of course, and the by-laws of Hull said that they had to be back over either North Bridge or Drypool Bridge before six o'clock. So they had to be out of Hull city centre and down Holderness Road, and they used to take these fish heads and spread it on fallows and then that was their day's work; put the 'osses away and they'd finished for the day.

In Holderness, two horses were usually enough to pull a waggon, even when taking a load of corn to the millers, so unless one of the horses was a check horse, a single line from the outside bit rings of each horse was sufficient to drive the team. When going to town with a pair of horses, some horselads put the halter shanks through a hame ring and tied the ends in a *swagger ball*, and had special leather straps instead of the crossed halter shanks which buckled to the hame rings, but instead of being fastened to the halters, they were buckled onto the bit rings.

Though relatively rare in Holderness, there were occasions when three or four horses were needed in a waggon, but a team of four horses was common when pulling a binder. On the Wolds by contrast, three or four horses on a waggon was a common hitch, as illustrated by a comment made by a corn merchant who saw Ron driving three horses in the snow of 1947.

> There was a feller from Maddocks and Marlows; that was a big corn merchant's at that time, and he said, 'I see you've got a lad off the Wolds then, Foreman, this year,' because three horses wasn't typical in Holderness. But on the Wolds, they had to harvest wi' three. On

the high Wolds they used to go and get a part load of sheaves at the bottom, pull halfway up and loaden again and then finish off at the top, and they had the three horses all harvest.

When driving four horses in a waggon, there were variations in how the reins were set up. When none of the horses wore a check and the horseman was standing in the waggon, each horse was tied to its neighbour by its halter shank, with a side string to the outside of each horse. (See Figure 3.)

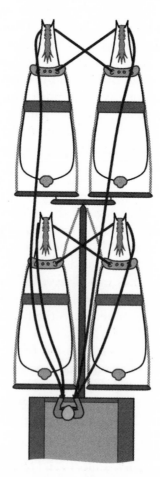

Figure 3   How the reins were set up with four horses, none of them check horses, when driven from the waggon. All the horses have a side string and are tied back to their neighbour with their halter shank.

When driving the same horses while riding the nearside wheeler, which was how a binder was driven, the lead horses were driven with crossed halter shanks and side strings. The wheelers also had their halter shanks crossed, and often had a single side string going from the outside of the nearside horse to the horseman, and to the outside of the offside horse. Alternatively, if the horse being ridden was a good one, it could be controlled using its hame rein, in which case a short side string went to the offside horse. (See Figure 4.) With a check

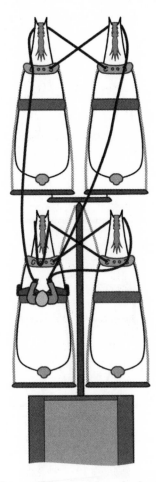

Figure 4   When driven from a riding position, the leaders' reins are the same as in Figure 3. In this case the nearside wheeler is steered with his bearing rein, while a short side string goes to the offside wheeler to keep him out away from the pole.

horse as a *leader*, it had its check rein back to the horseman, while the offside leader had a side string and was tied in to the check horse with its halter shank. If the nearside wheeler was also a check horse, a similar arrangement of reins was used on the wheelers, but if the nearside wheeler was not a check horse, the side strings and halter shanks were set up in the usual fashion. (See Figure 5.) When riding with a check horse in the lead, all the lines were just taken up shorter, or the wheelers could have a complete side string. (See Figure 6.) Alternatively, the wheelers could be driven as the wheelers in Figure 4. When changing between riding in the waggon or riding a horse, there was

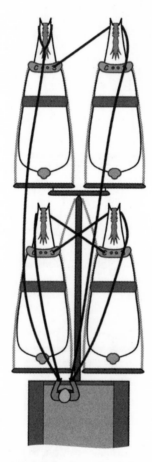

Figure 5   When the nearside leader is a check horse, the offside wheeler is controlled with a side string and is tied with its halter shank to the check horse's hame ring.

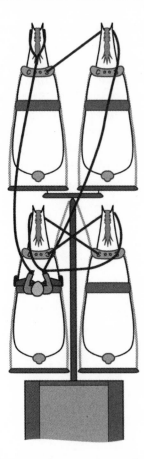

Figure 6 When driving from a riding position with a check horse in the lead, the reins are set up like those in Figure 5, or, in this case, the wheelers are controlled with a short continuous rein from one horse to the other.

often a length of cord to be taken up. The extra length could be wrapped onto the top of the hames of the horse being ridden, so if the lad dropped a rein it was still there.

Although details of the way horses were driven in Holderness are at variance with the rest of the country, in part perhaps because of its geographical isolation, there were in fact many variations between individual farms in the area. The divergence of opinion about check reins is the most obvious case, some horsemen swearing by them, while others swore at them. There were also other variables, such as where the side strings were tied, which ring they ran through, and whether the middle horse or the offside horse of a three-horse team had a side string. The variation in this type of detail, which now can only usually be discovered by examining old photographs, was the result of personal preference, the policy on a particular farm, or simply what equipment was available. Another variable was the individual horse. Some were happy to do anything and go anywhere, whereas others might go better on one side of a team, or perhaps be less trouble with a false line. Working out which arrangement worked best with each team of horses, in each particular situation, was part of the skill of the horseman.

CHAPTER 8

Breaking In

WHEN farms relied on horses to do the work, most farms bred foals, not just to replace the older horses when they got too old to work, but also to supply horses to do delivery work in the towns. For some farmers this was an occasional extra bit of income, but others specialised in supplying horses. Indeed, H Caley was an agent for the London and North Eastern Railway, who used large numbers of horses to deliver heavy goods on the last part of their journey from the railway stations to their final destination. Although the demand for town horses had declined because of the competition from motor lorries, in the 1940s the Caleys were still buying young horses which would be broken in and worked on the Caleys' farms before being sold in the towns.

On many farms the foreman broke in all the young horses himself, but others entrusted the job to a labourer or the waggoner. On the Caley farms, the horses were left completely untouched as foals, except on the rare occasion they were to be shown in their first year, in which case they were halter broken. All the other foals were turned away after being weaned until they were two or three years old, when they were broken in. Different horsemen had their own methods and techniques for breaking in, and much of it depended on the skill of the individual horseman. Nevertheless there were similarities between the horsemen, so although the method Ron describes is the way he learnt from George Gibson, on the neighbouring farms similar methods were used.

The training of the horses at Carr Farm mirrored the way everything else was done, every job being carried out in a specified manner, and in an allotted time. So the breaking of horses was done almost to a timetable, and the horses were pushed through the system until they could be used for work. The early training, when the horse was getting used to the harness and the bit, was all done in the stable, and there was a minimum of work done outside before the young horse was put to work alongside an older horse. But before anything else could be done, the young horses which had been left to their own devices for two or three years, first needed to have halters put on them.

> We'd get the young 'osses up when we'd got there at Mart'mas, and either put 'em in a yard if they hadn't enough bullocks, or put 'em in boxes. Some of these big boxes they'd put two or three young 'osses in. Us lads had nothing to do with 'em when they were in boxes, either Foreman or ' labourers fed 'em.
>
> Then when it come to ' end of February, they'd say, 'Right, we'll break some in,' so you'd have to helter 'em. Now they'd never had helters on in their lives, and Foreman used to say, 'It's no good us breaking these horses in if they aren't full of corn.' So they used to corn 'em up, to have 'em right at tip top, 'cause George Gibson used to reckon that if you broke a horse in that wasn't fit, it was coming to it before it was ready. Then you had to get some helters on 'em, and left 'em with their shanks hanging. Now in a box, if one 'oss didn't tread on the helter shank, the other one would.

Having the halter shanks trailing on the ground so the horses would stand on them started to teach the horses that they had to stop when they felt the pressure on their halters, which helped for the next stage, which was to teach them that they could not get away when tied up, swinging as it was called.

> After they'd been in these boxes two or three days they'd say, 'Right, we'll have 'em.' So they'd attach two waggon ropes, one to each helter shank, and then they'd have two or three fellers hold of each waggon rope, and make 'em get out their boxes. Once they'd got 'em out, three would be the anchor, and three would lead it, but it always wasn't the same three. They'd get it out into ' stackyard, across the road, and down to the other end of ' 'oss pasture, and at the other end of ' 'oss pasture was a post, and this post would be about six foot out ' the ground. Down they would go with this horse and get it to the post, both wind it round the post, and then they would set it off running. They would have a whip and a stick and they would bang on this rope, and the horse would be running round with its nose facing the post, running sideways right round. And the men, which would be Foreman and Boss, would be banging

on the rope with a stick, and every now and again they would just give his nose a tap. Well, every time they tapped his nose he used to fly back and swing his head, and they'd run him round 'til they'd got him wound right up to the post, and once they'd wound him up, then they'd run him out again.

While this was happening, the other men went back to the farm to bring another young horse, by which time the horse being swung would be white with sweat. The horses were then swapped over and the process started again, and if there were enough young horses this would continue all afternoon. After each horse had been swung it was returned to its loose box, still with its halter shank trailing.

Now, the next morning, they'd say, 'Tek that horse for a walk!' Two of you would just get him, and out he would come; out ' stackyard and down the road, and they took him so far and come back, put him away and got the next one out. They took them out every day, 'til about the third day you was on your own and he walked like a little lamb.

Then they took 'em in the stalls and tied 'em up. Foreman'd maybe be doing this and one of the labourers. This would be another day, when the other horses had gone to work. When they'd settled down, they put 'em in backwards. They were backed into the crib, and you tied one of the helter shanks round one pillar and the other to the other pillar, so it was stood in this stall pillar reined up. Then they could do what the hell they liked, they couldn't go forward, they couldn't go back 'cause of the crib, and they couldn't swing sideways. Then once they'd got 'em in backwards, they'd put a bridle on them and put a bit in their mouths, and he would stand in there for probably two hours, and they would drop their bit out and then take the bridle off and put 'em back in their boxes, and that was it for that day, making sure there was no 'oss muck in the crib, 'cause when the others come back, you had your *name for nowt* if you didn't have the cribs cleaned out.

Then the following day they would do exactly the same, but this time they would stop in a bit longer. Then about the third day he would think it was time they had a surcingle and a crupper. So he would put a crupper on and tighten the surcingle up, with a bridle and the bit in, and he would stand in there for maybe two or three hours, still backwards. Then after they'd been in there maybe a week, just Monday to Friday, the following week he'd maybe think they wanted a rein on. So there was some reins attached to the surcingle and they would put them to the bit loosely, and each day they put 'em in 'til they got them quite tight, but never no more than four hours. So he was stood there with just a little bit of tension on his bit, and this would go on for at least three weeks.

Once the horse was used to the harness and a bit, they were long-reined. If the foreman was breaking in a horse, this would happen during working hours, but if one of the horselads was breaking in a horse, they would do it after tea and were paid an agreed amount for each horse.

You long-reined each horse for an hour a day, and by this time you could slip the harness on quite easy, they'd been having it on for twenty days. One man would be driving them and the other man would have a line on 'em, into his bit, and the following week you would bring them out and you'd have a log or a gate post and you would hang them in the gate post. Then we would set 'em off, and some of 'em would walk away with hardly any problem, just jump a bit, and each horse would get an hour in the log.

After the horse had pulled the log or gate post for an hour every day, spring would be well underway, and the horselads were spending much of their time harrowing. This was the first time the young horses would do some real work. When starting off a young horse, the two older horses were first yoked together and to the harrow, and then the foreman would bring the young horse alongside them, and hitch it up on the offside of the team.

The first thing you did with any young horse was you yoked them in the middle of the field and set 'em off: ' didn't matter where they were going as long as they were going forward, and you could always bring 'em round in an arc. Then once you'd gone somewhere with 'em and he had settled down, then you could start and steer 'em, and you harrowed as though you was three just ordinary horses, and Foreman would go home. Or there may be two of you with a horse apiece, not necessarily in the same field, two lots of three horses. By about ten o'clockish these horses were beat. He would go home and bring some more. He would then tie one up and bring the other horse to you, and he would get one out and put the other one in, and you always put the young 'oss in last, and you took it out first, but it was still tied to the other horse's collar. Or if necessary he would bring someone with him, if he thought they were going to have a lot of trouble. Then he [that horse] would go while dinnertime, and that was them two horses done for the day, and they were put back in the box. But by this time you didn't have their helters trailing, you could slip their helters off.

Usually when starting a young horse in harrows, the two older horses were driven in as a pair in the normal way, and the young horse was then tied with its halter shank to the older horse's hame ring, with a false line running back from its bit rings to the middle horse's trace, and then its own traces were hooked onto the swingle-tree.

> That was the first thing you did when you got these young osses, put 'em at the off end with a false line on 'em; you'd no string on 'em, they could do the hell they liked. They couldn't go forrard with a false line on 'em, they couldn't go back because of the helter. But you'd no control over it; you just drove the other two. It would be prancing and dancing, and after they'd been in there two hours you couldn't see 'em, they were white over, 'cause they'd pratted about.
>
> But we always considered that a young horse, if you put him on a false line, he'd no weight on his mouth whatsoever, only what he put on himself. So if he wanted the pressure on he would go forward, but if he walked comfortable the line didn't hurt him, so he was doing exactly to his mouth what he wanted, and this was considered the way to do it.

After the horses had spent a week harrowing for two hours a day, the following week they would work all morning, and by the third week they would go all day, but by this time they might be pulling lighter straight-toothed harrows rather than the heavy gibb harrows. They would then be introduced to working in a pole.

> After they'd been going about three weeks, Foreman would say, 'Take a waggon down!' So you would take a waggon down with the older horses and tie the young horse behind, and if necessary you'd tie him with two shanks, one to each *airbreed*.
>
> Now when he came at five o'clock, you'd tie your middle horse at ' back of the waggon. Then you would put your young horse and your nearside horse in as a pair. Sometimes it was a bit of a struggle to get 'em square, and then he would have another string on your young horse, you would get in your waggon and off you go home.

He would walk, and he would let it have it, give it some string, and get it to jump and get everything to rattle. You might be three fields away from home, and ' course by ' time you got up to stable door, he'd be quite used to it.

When training a young horse it was important to have reliable older horse to work alongside, especially when putting the young horse in a waggon for the first time, because the empty waggon was easy to pull and made a lot of noise as it rattled and the iron tyres crunched on stones.

They were very particular which horse it was, it hadn't to be any horse, it had to be a very good horse what wouldn't tek fright, because these horses when they got in that pole they'd think nowt of jumping over the pole, and you'd have two horses at one side of the pole. You'd be in the waggon and the foreman'd have a string on the young horse and he'd say, 'Keep 'em going.' Then when he thought they'd gone long enough he'd say, 'Just stop,' and we'd get it out, and get it to ' right side again and he'd say, 'Right, off again,' and he'd let into it with his string and mek it jump a bit, and eventually he would get in ' waggon and we'd go out the field and across the road and round the stackyard, and he'd say, 'That'll do,' and we'd louse 'em out and put 'em in ' stable.

Then the next morning the young 'oss went straight in the waggon, and when he got in there didn't he let him have it. Oh hell, he'd strike him, and the waggon'd rattle and the 'oss'd be jumping and, 'He'll be alright,' he used to laugh, and that was how he went on. We'd take 'em back where we was harrowing, then louse 'em out, leave the waggon there and put 'em in your harrows. Then he would come again at dinnertime, and take 'em back, and they'd been in the pole then, that would suffice.

Once the young horses were working regularly and were being driven with a side string, rather than being controlled with a false line, having the cord rubbing the underside of the jaw could cause it to become sore.

We always drove them with the strings under their jaws, and after you'd had them out for probably a week, under their jaw got quite red raw. So you used to put your string through the nearside bit ring, tie half a bowline, put it through the farside bit ring, brought them through and continue with your bowline. So when you pulled, it was pulling at both rings, but no more was it touching the jaw, so the horse had no discomfort. If this went on, we used to just gently wash them and then rub their jaws with methylated spirits at night, and you drove 'em with your string not touching their jaws 'til the jaw had got nicely healed and then if it occurred again, you just put your rein back to where it was. All horses didn't get sore but some had more sensitive skins than others.

Up until this stage in the process the young horse wore an open bridle, but by the end of spring the open bridle was replaced by a blinkered bridle.

Then the next thing that come along would be muck leading. So these young 'osses went straight in the pole with an old horse and you got muck out. Then you'd bit of a pantomime 'cause they knew they were going to have to pull, and they didn't always take kindly when they hit their collars and the waggon didn't move. But if you'd got the good old 'oss, he would pull, and just a little persuasion the other 'oss would go.

Leading muck from the different fold yards often took a month, so by the time all the muck was out, the young horses were well accustomed to the pole and pulling a waggon. The young horses also pulled the waggons at hay time and harvest, and then got plenty of experience in chains when ploughing and harrowing in the autumn.

Being yoked in front of another horse for the first time did not happen until the potatoes or mangels were lifted, the young horses being put in first horse gears in front of an older horse in a cart. By this time, the young horse had done everything except been put in shafts, so he still needed to get used to pushing the shafts around when turning, and pushing into his britchin to back the cart. This was also usually done in the mangel field, after the horse had become accustomed to the routine as a trace horse, except on the odd occasion where an exceptional horse might have been used between the shafts of a fertiliser drill.

Once harvesting the mangels was finished the young horse had been used with all the different types of harness, and with many different vehicles and implements. Over the following months and years his daily work gave him more experience, teaching him to accept new sensations, such as the vibration of a mower. For some horses, especially the mares, the yearly cycle of work on the farm would continue for the rest of their lives, but for many of the geldings a year or two of farm work was just a temporary phase, before they were sold to pound the hard streets of the towns.

CHAPTER 9

Marton

BESIDES breaking in the horses bred on the farm, the farms run by the Caleys also broke in horses that were bought in. Many were bought locally and most were easy to deal with, but some presented a greater challenge, such as a pair John Caley had bought before Ron went to Carr Farm.

They'd had a pair of horses, two-year-olds from Scotland. They were slate grey with white legs and manes, big 'osses, and [one of them], ' first thing he did was jump out the lorry and he smashed somebody here. They went for the other one, and it hit him fair on the head. This is before I went. So they give 'em about six or eight weeks work and turned them away, and Wag hadn't been very pleased, he was a bit frightened on 'em.

Anyway, so we gets there, and Foreman says, 'We'll just go and fetch them 'osses. Get your bikes!' They were turned out were these horses, and there was a racehorse with them; its mother had won one of these big classic races, and this foal had been fetched up on the bottle. So down the road we goes, me and Third Lad. We gets to this gate and he whistles and they come, he puts a helter on this blood 'oss and he jumps on its back and they'd gone; full hell down ' road, with these 'osses at back of 'em. So we got Foreman's bike and our bikes, and after 'em, and when they got there, they were waiting. We'd come three-quarters of a mile down ' road, turned in at Carr Farm, and we got 'em in a corner and he said, 'We'll 'a' some helters on them,' and they had a struggle.

And Wag says, 'Don't put 'em in my end, Foreman!'

So he says, 'What do you mean?'

'I want nowt to do with the buggers.'

'Oh alright,' he says, 'put 'em at Thoddy's end!' So we fastened them up in there and he says, 'Don't touch 'em, don't try and muck 'em out! I'll come and feed 'em.' He come in a morning, cut hisself a lump of wood, a point on one end, and he goes up [between the horse and the standing] and as soon as he gets up, it swings, and he just stands still and puts his wood out, and one end's on the standing and the other end's in his ribs, and it jumped up. As soon as it jumped up he went in, fed it, got his stick again and he come out. He said, 'Don't you go up there! I'll come out at dinnertime;' never got no water. After he'd had his dinner, he goes up with his stick again. By, it come again, and he stuck it there and stood still. He got hold of it and took it into ' fold yard for a drink and a sparrow just flew over, and it went up in the air, and ooh hell, it fought! He brought it back; same with the other one. We'd put some grub in while it was away. He said, 'Don't you touch 'em!' He did that for a week. He says, 'There, they'll be alright now.'

Well, Thod Lad had to look after 'em, and by this time, for some reason Wag had fallen out with him, and he left. Now if you're hired by the year, usually you leave without your money, and they won't sack you within the year either. But things were changing. Anyway, Wag left, they wa'n't going to get anybody else, so I took five over and Thod Lad had five.

The departure of the waggoner meant that the third lad moved up to being Waggoner, and Ron became Third Lad. This situation continued until the following Martinmas when Ron had to decide whether to stay on for another year. Although it was customary for horselads to move to another farm every year, by this time many farmers were using their horses more for odd jobs, rather than keeping them in full work in the traditional manner, so both the new waggoner and Ron decided to stay with John Caley for another year. But Ron's time as Third Lad was to be short-lived, as a few weeks later the new waggoner also left, so Ron became Waggoner, and had all the horses to feed and muck out. A few years earlier, if a horselad left his employer they would have found a replacement, but in 1948, the horses were no longer as essential to the farming operations, and also there were simply not enough lads wanting to work with horses to fill the vacant places. Because there was no one else working full time with the horses, Ron often worked on his own, and when going to plough he would take out a different pair each time, except the two Scottish horses, which were always worked separately along-side a quieter horse.

I was going to plough every day. I mean, by now we were only ploughing out of courtesy; there was crawlers and I don't know what we hadn't. And Foreman come to me one day and he said, 'You know this is about the end of the horse era. I've spent some of the happiest days of my life ploughing.' He says, 'We'll show you a few tricks,' and then he'd plough a piece, and he'd put all the *garings* in there was, and then he would show me how to take them all out at one go, and he'd show me how to plough square corners, and he showed me all sorts of dodges. Oh, he was real keen to show me because he knew this was the end. I mean, had it been fifteen years earlier he wouldn't have shown you, you'd have had to fathom it all out for yourself. And then come the spring he says, 'Have you ever used a rowing plough?'

I says, 'Not much.'

'Well, I want these taties banking up. There's ten acres over there. I'm going out.'

I said, 'Alright,' and I took one of these Scottish horses with one of my other horses and we were going away quite nicely, and then there was a little paddock right near stable door and I thought, 'There's nobody about, I'll take 'em both.' I'd worked 'em both as individuals or with one of the other horses. They'd now been shafted, and I'd had 'em in a waggon, and I got so I could walk in ' front legs when they was in ' stable and come out their back legs, and they never as much as flinched. So I put 'em in this rowing plough and every time we come to ' end where ' stable door was, they wouldn't go, and we had a hell of a do. I'd come out of one row and I maybe had to miss four before I got down another one, and I'd to match 'em up as I was going round best I could. I was having a bit of a pantomime, and they'd set off to back and I could pull their tails and I'd still hold o' the plough hales, and they would back over me if I didn't watch them. And just as I was having one of these panto-mimes, Foreman come.

'Now then,' he says, 'what are you up to?'

I says, 'I thought I'd give 'em a go on their own, Foreman.'

'Aye,' he says, 'you know what you've done wrong, don't you? You've picked ' wrong field. You're too near home.' So he led 'em off for me and walked up and down with me, and he says, 'You're alright, you keep going!'

So in the afternoon I got 'em out and he went and fetched John Caley, and they come to have a look at his horses going as a pair.

Oh, I did all sorts with them, and then I left that Mart'mas, and he says, 'You aren't going to stop, take them two and shove them in that box!' And I shoved 'em in a box, and I went from there to Marton.

When Ron had been asked if he was going to stop again in the weeks before Martinmas, he was unsure whether he should stay at Carr Farm for another year. Even though John Caley was keen to keep him, during the conversation on Martinmas day Ron made up his mind to leave, so once again he found himself standing outside Holy Trinity in Hull on Martinmas Tuesday.

We were walking about and up comes John Caley, who I'd just left, and he says, 'Now Ron, 'a' you got hired?'
'No.'
'Come and have a drink wi' me!' So we went in ' Corn Exchange. In them days there was all sawdust and spittoons in all the pubs and he was talking to me, and he had me waiting there while his brother, Norman, come.

Norman Caley was the youngest of the Caley brothers, who farmed at West Newton, a village a mile or so north of Flinton, but he also ran some other farms, including Lane End Farm at Marton where his father, Harry Caley, had started his farming career fifty years earlier.

So Norman comes and he said, 'Now then, 'a' you got hired?'
'No.'
'You're going to come to me, aren't you?'
'Well', I said, 'last time I was at your place, the bloody stackyard was mud up to your eyes,' because when I was at Flinton I'd had to go and borrow so many carts from West Newton.
'Well, we'd been wuzzeling up and down it, no doubt it is mucky.'
'No, I'm not coming there.'
'Well, how about going to Marton?' he said, 'Briggs is a good feller; been with us twenty-five years, has Briggs.'

Gil Briggs, the foreman at Marton was a similar age to George Gibson, but unlike George Gibson and many other foremen, he had stayed on the same farm, and it was with him that the Caley boys had gone to work as young men in order to learn the ropes before taking on farms of their own. Being well established, Gil Briggs had more influence over the running of the farm than some of the other foremen, so for instance the horses were fed hay, which was unusual on farms in the area. For Ron, a major attraction of going to Marton was that was one of the few farms run solely with horses.

> So I decided, 'Alright, I'll go to Briggs's.' I got me five shillings fest, and instead of twelve pound ten for doing horses, I'd twenty-five pound, because I had to help 'em wi' bullocks on a Sunday morning.

At Lane End Farm there was stabling for five horses, and in the past there had been work for two horselads, but the number of horses had now been reduced to three so Ron was going to be on his own. The exchange of information between the lads at the hiring fair, however, did give Ron an idea of what he might expect the following week.

> Walking about I met a lad. I said, 'Are you hired?'
> 'No, whe're you going?'
> 'Briggs's.'
> 'Oh', he says, 'I've been there, plenty to eat.' When you met anybody that was the conversation, how good a place it was to eat, 'cause some of these places were a bit rough. He said, 'There's only one thing, she's allus slow getting the dinner ready,' because when it was dinnertime you got your dinner, didn't matter if was ready or not. You went in at twelve o 'clock; you went in full hell and you got your dinner, but the taties weren't allus ready, but nobody commented on nothing.
> Anyway, so I was going to Norman's, and what Norman says was, 'He's got three daughters, has Foreman, and be careful what you do with the middle one, because she's the apple of her father's eye!' This was the Tuesday, and I was going there the following Sunday night, and when I got there, Briggs says, 'We were just going to bed,' you're late, sort of thing, 'have you had your supper?'
> 'Yes, aye.'
> 'Well,' he says, 'you don't get none here,' and he meant it, an' all! Your tea was your last meal in any of these houses.

Although the domestic arrangements were similar on different farms, a lad would not know quite what to expect from the foreman or his family when he went to the new farm. Usually there would be other lads to talk to, but being the only horselad at Marton, Ron just had to work things out as he went along.

> He says, 'There's three 'osses and a cow in ' stable. You know where stable is,' and sort of just pointed out there. So I thought, 'Hell, now what have I come to?' you get no supper here, sort of thing. Anyway, in that Mart'mas week I'd met the kid who'd been there the year before, and he'd said there was a cow in ' stable and all this!

On most of the farms the foreman had the use of a cow, which he had to feed and milk himself, and if the farmer lived on the premises there would often be another cow for his family, but this cow was milked by the bullocky. The foreman's cow provided milk for the foreman's house, including the lads, and was one of the perks of the foreman's job. Keeping it in the stable however was unusual, as the stables were built to house the right number of horses to do the work on the farm. Ron's surprise at finding a cow in the stable is explained by the fact that horselads, except on occasion the least lad, were hired to work only with horses; the other livestock usually was nothing to do with them.

> Next morning I went and fed these 'osses, turned 'em out, all three at once, and they went for a drink. While they was out I mucked 'em out, I thought, 'Well, while I'm on, I'll muck the cow out!' It was straight opposite the door, and I mucked the cow out and then brushed these 'osses, and he didn't say owt about gearing 'em, so I didn't put no collars on, and he blew a whistle. When I went in for me breakfast, there was this girl. I didn't know who the hell she was. I thought, 'Well, she isn't old enough to be his wife,' and she was his eldest daughter. There was these three lasses at home, and I thought, 'Well, where the hell's Missis?' We had our breakfast, and at all these places they used to lay a knife on the pie and you cut your own piece. I had my bacon; just fat bacon and a pint of tea, just the same, and I saw this pie so I did no more than just got it. I didn't wait for him, and I cut a lump of pie, and he says, 'You haven't cut that pie, Lena!' And Lena should have cut the pie. I got this pie and there was no other! And I thought, 'Just one pie?' So I just got up again, and cut myself another slice, and nobody said nowt. It was apple pie, and I thought, 'Well, I won't be greedy, I'll just have two lumps.'
> And all the pie there, they had a big orchard, so there you didn't get many prunes and figs, you got a lot of apple pie.

Understanding how things worked in a new place was something a lad would soon learn during the first few days, but Norman Caley's advice about the pretty middle daughter, became clear as soon as Ron saw her for the first time that afternoon.

> When we comes in for tea, there was the full contingent, the three girls, and by, when you saw middle girl, you knew what Boss was talking about then, about not getting her into trouble! Anyway I went out every night doing these horses and in comes this girl, and she would brush 'em, and as the nights got lighter she used to come down every night to where I was ploughing. I used to give her a leg on, and she used to ride one horse back, and when I got back all the horses were bedded up, all the racks were full of hay, and all I had to do was to feed 'em, brush the stable up, and brush the horses, and she used to come brush 'em.

Being the only outsider in the foreman's house, and with the foreman's three daughters living at home, meant that the dynamic in the house was different from when Ron was one of three lads living in the foreman's house at Flinton. Usually the lads were kept at a distance; they were there to work, not to become part of the family, so Gil Briggs asking Ron to sit in the house on the second night, and the conversation that followed, was unusual.

> The following night when I went in, he said, 'You can come and sit in the house again' the fire.' Well, none of the others told you, you could sit in the house. Some of them, they'd only invite you in [occasionally]. You had to sit in stable mainly, and he says, 'What have you done with your money?'
> 'Oh,' I says, 'spent some.'
> He says, 'I hope you've saved some,' and he says, 'once a man and twice a child: when you're young you've got your mother to look after you, but when you get old you'll want your money to look after you.' That was his criteria right through his life, and he says, 'When you've had one pint of beer, you're never dry no more, so you don't need two.' And he used to go up to ' pub on a Saturday night and he used to follow that principle. One pint would do him.

Gil Briggs's careful ways, besides limiting the amount of alcohol he drank, had also enabled him to buy a car, which was a rarity among farm workers, even the foremen.

He was the only one that had a car, an old Morris or something like that, and they used go out all over, we used to all go to whist drives in it, and one night the girl, this girl about fifteen, she was the apple of his eye, she couldn't do no wrong, she rushed out and got in the car and off she went down this lane, onto the road and up another lane and he played hell, and she eventually came back with the car, but he didn't half tell her off. But all she'd do was laugh!

And they used to go off in car at night and he used to say, 'You needen't go out, you can allus sit again' ' fire,' and so when you went in you could have a chair there but he never offered you anything to eat, and some nights he went out and there were these lasses in, and oh hell, we used to have hell on earth in there, like, ever such fun.

The pattern of work at Lane End Farm followed a similar pattern to Carr Farm, but because there were only two of them working on the farm, although Gil Briggs did most of the work that didn't involve the horses, Ron was still needed to feed the bullocks during the week and at the weekends as well.

All these bullocks were fat bullocks so we used to cut [mangels] for them, and we had an engine, and he says, 'You can fill ' turnip cutter up.' So I was putting them in this thing, and ' fast as I could get 'em in, it was cutting 'em and he was shovelling 'em away, and he said to me, 'When you're putting them wuzzels in that cutter, I'll give you half a crown if you can bung me up.' So after about five days he says, 'I'll give you a swap, you can shovel!' And when you were shovelling out, you had to spread them out on this chaff, and then you'd put some more chaff on, and then have another go, and it'd last you 'til the next morning. So he started, and I realised then what I'd been doing; hell, I didn't half have to shovel, and he knew an' all. He says, 'Now then, what do you think to that, then?' And I knew full well that I needn't go as fast as I had been doing. I went a lot steadier after that.

Because of the work with the bullocks, Ron did not get away to plough until halfway through the morning, and the very first time he did any ploughing, Briggs made sure he went with him.

There was about seventeen acres of kale land to plough, and he said, 'I'll come with you and help you set ' riggs,' because he wanted them straight, right on ' roadside it was. I was ploughing away, and he was hedging; and all these hedges, he had a lovely little slash knife and he'd just twig them.

The concern to make sure Ron was up to the job was partly because good ploughing was important to the effective cultivation of the fields, but being on the roadside, the ploughed land would also be seen by all the neighbours in the area, and a badly ploughed field would reflect badly on the foreman and the farmer. Having spent the mornings ploughing, the horses returned to the stable at dinnertime, and then after dinner the bullocks were fed again before Ron returned to ploughing. Instead of carrying the feed in a sheet, at Marton the feed was carried to the tumbrils in big baskets, and then in the evening the tumbrils were filled with hay. Before baling was common, the hay which had been stacked loose in haystacks had to be cut with a hay knife into manageable-sized pieces. The front of the stack was cut into three divisions, each section being referred to as a *dess*.

> At night he would carry all this hay into these bullocks and when I come up, he'd done the hay. Then you had to do your horses at weekends: there was no weekends off. On a Saturday morning we carried enough hay into a barn in these big desses, and he would carry them out himself on a Saturday night and on a Sunday night. And on a Sunday morning we used to feed the bullocks once, then go and have our breakfast, and then we used to feed them all again. Then very often he would say on a Sunday afternoon, 'I'll do 'osses for you tonight,' and you'd go off after dinner.

In the spring of 1949 there was a big sale at West Newton as the Caley family reorganised the farms after the death of their father, and equipment was brought in to the sale from a number of the brothers' farms.

Farm sales are always a highlight of the farming year for farmers in the district, a social occasion as well as the chance to buy farm equipment, but the sales at this time also ushered in a change to tractors, so much of the horse-drawn machinery was auctioned off for whatever money it would make.

> When I was at Marton, Foreman says, 'You've to go to ' blacksmith's shop and get a *breast pole* for a corn drill.' So I went to ' Withernwick blacksmith, what Marton used, and he said, 'While you're going, you might as well take the bill,' and he wrote the bill out for seven and sixpence. So I come back and we put the breast pole back on the corn drill. I'd to take this corn drill eventually to put

it in the sale, and when they sold it, the drill fetched seven and six, and we'd just paid seven and six for the breast pole!

The sale also saw a reduction in the number of horses, and although there was no future for the old and the young unbroken horses, there was still a market for work horses, including the two which had been brought from Scotland, neither of which had done any work since Ron left Carr Farm the previous Martinmas.

All the old horses went for slaughter and all the young horses went for slaughter, and twenty-one horses was going to be sold as sound working horses. They came in from Aldborough and the Fossams, and these two horses from Flinton came, the two young grey horses. And of course they fetched all brand-new helters out for 'em and all the ribbon, and when sale day came, the two main fellers plaited all their tails up and they all had white and blue ribbon in their tails. And there was bullocks, and all the implements and all the sheep and pigs, all going to be sold. All the horses were sold, and the top horse

Advertisement for the sale at West Newton.

made seventy guineas, and these two horses from Flinton which had cost a hundred pounds each about three years previous, the top horse made forty-five guineas! One went to ' railway in York and where the other one went I don't know.

The low prices achieved for the horses reflect the fact that fewer people were using them, the late 1940s seeing the fastest reduction in the working horse population. With high prices for agricultural products, farmers had money to spend on tractors and were keen to produce as much as they could, so the use of artificial fertilisers also increased. At this time the first weedkillers also came into common use, albeit with very little regard for their harmful effects, as Ron found out the first time he came into contact with them.

There was a trip going from ' Bluebell pub at Sproatley to Whitby, and I hadn't put me name down to go on this trip, but on the Thursday night someone comes along and said, 'There's a ticket for you, Ron, if you want to go.'

So Friday dinnertime come, and I says to Foreman, 'Will it be alright if I go to Whitby tomorrow?'

He said, 'Well, if you drill that nineteen acres.' We were putting weedkiller on; it was first time we'd put weedkiller on in a powder form and it killed charlock, or runch as we called it. So he says, 'I'll bring you another 'oss out at five o'clock.' So I set off at one o'clock, and you only was putting 200 weight to the acre on, and he come out at about half past four with another 'oss, and it'd be eightish and I'd drilled this nineteen acre of peas. And while I kept filling this drill up, some of it had got down my sock top, and I didn't bother, I was wanting to get this nineteen acres drilled. And, by, when I got up on Saturday morning, I'd burnt all me heel with this weed killer. But we went to Whitby on the Saturday morning, but I had quite a bad heel for a day or two.

Having a day off work was a rare occurrence for a horselad, but from 1938 the lads all had a week's holiday, which was often taken just before harvest. With livestock, however, the need for them to be checked every day meant that Ron had to take over looking after the cattle in the flood meadows when the foreman went on holiday.

Summertime every morning, he'd say, 'I'm just going round bullocks,' and I, whatever I was doing, I was left on me own. And one day, getting near ' holidays, he said, 'I'm going on me holidays, such and such a week. I'm going to Blackpool,' and of course I was having my holidays an' all; I think we were going to Butlins. So he said, 'I'll take you down to Lamb'aths and show you how to carry on.' Well, ' Lambwath is a stream what starts at Aldborough and it wends its way right through all the villages, and it drains a lot of the land in that part of Holderness, and it lands up in one of the drains and eventually into the river Hull. And all these Lamb'aths would flood, all winter they would flood, and every farm on that side of the Burton Constable estate had a Lamb'ath. They were all in ten-acre pieces and there was no hedges, there was side dykes. And Caley's had their bullocks in because it adjoined the farm, so it was all fenced for bullocks. But a lot of 'em, all they did was make hay in it, and nothing was done with their Lamb'aths 'til the following hay time. And it was nothing for it to flood while you'd all this hay; you'd got it all cocked and it would flood, and then they had their boots off trying to float these haycocks to a dryer part, and when you went to these bullocks it was ankle deep in water, but ' course it made the grass grow.

So he takes me down, and there was a hundred bullocks and there was a row of tanks, and he says, 'You take a bucket and go down into this stream and prime your pump, and then you start pumping.' You had a big long piece of hosepipe, and you put it in one tank and you started to pump, and when you filled that one, all these bullocks were there drinking; drinking and fighting for water and he said, 'You pump 'til all these bullocks have walked away and all these tanks are level,' and I thought, 'hell, he's off on his holidays!' I realised it was no holiday when you got there: hell, they did drink.

By the time Ron had also been away on holiday, it was nearly harvest. Before the war, the farm at Marton always had two Irishmen to help harvest, and when they were leading sheaves they would all go to the field with two vehicles, load them up, then all return to the stackyard to team them, the whole process being repeated until the corn was all safely stacked. But with a reduced labour force and the new trailers which carried more than the pole waggons, the farm staff from different farms all worked together.

This is when they came in with all these Tye trailers, and we had two pair of horses, a pair in each trailer and cleaned that farm up, and then we went on to West Newton and High Fossam and Low Fossam. We had quite a long harvest, it'd be mid-July when we started among peas and and it'd be nearly into October when we finished.

If it had been possible Ron would have stayed with Briggs for another year, but Norman Caley had decided to give up his tenancy on Lane End Farm the following Lady Day, 25th of March, the traditional day for starting and ending tenancies. So while Gil Briggs stayed at Marton until the end of the tenancy, after Martinmas Ron started work at Old Farm, West Newton.

CHAPTER 10

West Newton

OLD Farm, West Newton was the farm where the Caley brothers had been brought up and where they worked until they were old enough to take on a farm of their own. Being the youngest, Norman Caley continued to farm there throughout his working life, but he was also responsible for adjacent farms, which were now being farmed as one piece.

I went there for waggoner, and we'd had the sale so there was now about seven horses left. But before I went, there was several horses at High Fossam, they had fourteen horses at Low Fossam, and nine horses up at West Newton, and when you went round the back at Low Fossam, you'd see sixty or seventy young 'osses all turned out, waiting to be broken in. Well, after this sale, it all altered. There was one at High Fossam and one at Low Fossam just for the stock man, or the bullocky as we called him, just to get straw and kale into ' bullocks, and the rest was all up at West Newton, and there was only just one foreman at West Newton and he would foreman the lot.

The foreman at West Newton, in stark contrast to George Gibson, was not at all interested in the horses, so Ron was left alone to do his work. Sometimes the bullocky lad drove a second team of horses, but otherwise Ron looked after all the horses himself, and although the horses were used less for the arable work, the morning routine remained the same.

It was the same procedure, a pair of horses in waggons, or pulling them swing. We pulled swing into the yards 'til the muck got up, 'cause it was downhill into the yards; they were all dug out, so when you were stood in some of these yards, I'd a job to get the *sneck* to open the stable doors. But by ' time spring come you was having to dig the doors out to get 'em to open.

In the eastern counties where arable farming predominated, keeping bullocks in yards over winter was one of the traditional ways of maintaining fertility. In East Yorkshire, as in other arable areas, there was a large quantity of straw to use, so the cattle were well bedded up. Although the yards were designed for this, a change in how the bullocks were kept meant the depth of muck increased.

> In the days when we had horns on the bullocks, there would be about forty in each yard. Four bullocks with big horns was all that you could get to a tumbril, and it was this feeding space that determined how many bullocks you had in a yard. But once we started cutting the horns off you could get eight at a tumbril, because [without horns] they couldn't fight each other. When you first went in these yards, we could ride in the waggons stood straight *over end*, and go in the sheds and come out again, but these yards were then doubled up to such an extent that when you were going in just before you'd turned ' bullocks out, all the hames used to catch the beams and the collars used to go flat down their backs and the horses were ducking. If you was in your waggon, you was right down in the bottom with just your eyes over the top, looking where you was going, else your head would be hitting these beams, and yet that was the same sheds you could ride in stood up.
>
> Once ' muck was high we had all sorts of problems. When it was spring, when these horses went for a drink, their feet were almost higher than the trough, and as soon as they put their head down, their collar went down to their ears and then they would be throwing their heads back. We'd a young horse one day and it put its head down, the collar shot off into the tank, and it flew back and ' stable door was open and the sneck stuck right into its hindquarters and split it open.

Because the depth of muck varied so much over the year, the big gates into the fold yard had to have more than one set of hooks to hang them on, so the gates could be lifted higher as the level of muck rose.

> All the gates across these yards had two or three sets of *crooks*, and the first week or two the gate was almost on the floor, and then it was lifted up; and when we'd turned ' bullocks out and got the muck out, you could walk down the yard and under these gates without ducking your head. Anybody not knowing, you'd say, 'What the hell was they keeping in there?' You'd think we had elephants, 'cause the

gates were so high. And of course, all the muck was got out with waggons and forks, and you could fill downhill into a rulley for the depth of the muck, and I've seen us, when we've pulled up ' side, a bullock walk off the muck and get stood on the rulley! And pigs come off and stand on the rulley! And if pigs got on we used to move with 'em on; they used to wonder what the hell had gone on!

'Cause in all these yards, not only was there bullocks but there was also all these bacon pigs; each yard would have about two hundred pigs. In the summertime they had the yards to theirselves, but when it was wintertime when you fed them, you shouted of 'em and they went up onto a causeway which was railing'd off. Then when they'd had their grub they went back into these yards, and anything these bullocks knocked over the sides the pigs would eat, and also they made all their beds in the straw. They always went under the sheds, of course, and in a morning when you were going in with these loads of straw, you went down the open yard, then you turned in and went down the sheds. Then you was among 'em and it was semi-dark, and there was such a squealing and a going-on, you'd pigs running all over, and these 'osses weren't a bit perturbed; they were quite used to it. They used to be struggling with their feet among all these pigs, and these pigs used to be running; and then the waggon of course; and only once we killed a pig. We run over it, and of course rationing was still on, so that was to the advantage; just took it in and dressed it, and ate that one.

Through the first winter at West Newton Ron was the only one going to plough with horses, in between threshing and riddling potatoes, but in the spring all the horses were called into action to harrow before and after the corn drills.

Although they'd have a tractor drilling, we would be following with three and four sets of three-horse harrows, and then I said to him, 'Why don't we have a four-horse set?' And he said he'd never seen a four-horse set going at West Newton, but down at Little Humber where they had 2,000 acres, they all drove in fours, but never at West Newton. But I drove in fours, and at one time I had four grey horses, which was quite unusual.

It was during that first spring when Ron was busy harrowing that one of the foreman's comments sparked Ron's competitive nature to show what he could do with horses.

138

Foreman said, 'Of course, when we were young, it was six 'til six.' That was agricultural hours. He says, 'You wouldn't stand that for long!'

So I says, 'Don't you worry about that,' so I said something to Norman Caley.

'Well,' he says, 'it's up to you. You can do what hours you like.' Because at one time of day when it got to be five o'clock with horses you stopped, unless they said, 'Well, you'll just run on a bit to finish?' And you might run on 'til six. Anyway, by now we were starting at twenty past seven, 'cause they brought the hours down an hour a week, but they wouldn't let you have the full hour so they took it off in ten minutes every day. So I used to go off three 'oss harrowing, and then come home at twelve o'clock, and went out again at one o'clock. Then I used to come home at five o'clock, feed 'em, have me tea, get 'em out again and go off again 'til as much light as I had. And so I was going from just gone seven 'til sort of nine. As the spring pulled out I wasn't always gibb harrowing; they'd maybe been drilling, they used to drill with tractors but they didn't harrow in. Then we used to harrow in, and you used to have six harrows, and maybe harrow twenty acres at night. And on a Saturday, when I went at one o'clock, I didn't bother to come home at five; I used to keep going 'til about seven or half past and then come home, and I did that for about six weeks.

And I was harrowing away one night, I'd harrowed this thirteen acres, and then I'd to go across with some straight-toothed harrows, and Foreman and Boss come along, and for what reason I don't know, they all laid down in ' middle of ' field. And Foreman says, 'You're overdoing 'em,' and I says, 'When you have a good feller you haven't any good 'osses!'

'Well, there look,' he said, and they all just went down, 'you're overdoing 'em!' Well, we wasn't really, because when they were going from six 'til six, they did two six-hour stints. Well, I was only doing twelve hours the same, but we were doing it a little bit different, and he said, 'We'll have some fresh 'osses for you.' So I took some fresh'ns out after tea. But I never did anything like that any more. I just proved to 'em that we was just as good a fellers as they were.

Despite the amount of work Ron got through with the horses, most of the arable work was being taken over by the tractors, and although at certain times of the year, when lifting potatoes and mangels, all the horses were fully at work, once the winter arrived it was no longer necessary for the horses to go ploughing. This meant that Ron's work was set to change, starting one morning when the foreman sent him to see Norman Caley.

One day Foreman said, 'Go round and see him!' So I go up to ' house, and he said, 'Have you ever fancied going round sheep?' And I'd seen this shepherd going round; well, it was chicken man who used to go round with a lovely strawberry roan pony and this little cart, and I used to think, 'that doesn't seem to be a bad job, they always seem to be going somewhere.'

So he says, 'Go and see Cliff, and tell him you're going to do ' sheep,' 'cause this pony used to take the water and the food round to these chickens as well. So I went and told him, and got the pony, and nobody said anything to me about this pony. All the sheep were in all the parks around Burton Constable [Hall]. So I goes off and I got to the first gateway and opened the gate; you used to open the gate and click the pony through, shout 'Whoa' and then shut the gate; and when I turned round I couldn't see the pony! The cart was there, but no pony! Then I realised the pony was laid down in the mud and he was trying to roll through this gateway. You only had to shout at him and he stood up, but by this time he was mud from his nose to his tail, 'cause in wintertime these gates were two foot in mud. So I went round and had a look at these sheep and went back home; the pony was allus kept in a different place to the big horses; put him away, and eventually I saw Boss and he says, 'Now then, was them sheep alright?'

I says, 'They was as far as I know, but I don't know what I went for!'

He says, 'Did you count 'em?'

'No.'

'Oh!'

I says, 'If I'm going again, I want somebody with me to tell me what I've got to do.'

'Oh well,' he says, 'you'll be alright. Would you like to go again tomorrow?'

I said, 'Aye.' So off I goes; nobody went with me and I rode round, and by ' time you've rode round two or three hundred acres it's a long way, and it was about a mile and half to get there.

Anyway, after a day or two Foreman says, 'When you've finished strawing, go round ' sheep!' So I said to Boss, 'You ought to come with me.'

'Oh alright,' he says, and he gets the pony and cart, and of course, when he got in he wanted to drive, and this was how he shepherded when he was young; they all shepherded with a pony and cart, and he took me round and he showed me how to count 'em and he said we would soon be feeding 'em. And we'd took enough sheep troughs out, over the different places I'd been at, so I knew how to put sheep troughs out. When the troughs had to be taken down, I'd loaden them on a waggon and rope 'em down, and then you'd lay 'em all out with about three foot between every trough, and when you first started to feed these sheep they got a quarter of a pound of grub each. All this grub was hand mixed up ' grainery and then bagged up, and he said, 'Whatever you do, make sure you get it in quick,' and by hell, you hadn't half to move. Else the big sheep, they were getting a mouthful, and they were following you and they'd knock you down, and when you were down they were all over you.

So you went down with your pony and cart and put a bag every three troughs, you tucked 'em in each trough before you started so that all the bags were out. Then you fastened your pony up to a four-stone weight on his helter so he couldn't clear off. Then you fed these sheep, and you fed 'em for six weeks before they lambed, all the lambing month and a fortnight after, so you was feeding 'em for about three months. Well on these Sunday mornings some of these lads said, 'We'll come with you to ' sheep if you like.' So they come with me and I says, 'Put all these bags out, and then you get hold of one of them bags and run down them troughs and trickle it out, but mind they don't knock you down!' Oh, they laughed! Anyway, once we started, and one on 'em, a sheep ran through his legs and it had him down. There was meal all over. 'Course, all the sheep rush to you, and you were laid there with maybe about fifty sheep trying to scrabble to get to it, and they got a bit of a shock did these lads. It happened to me several times, but when I got going I had a dog, and if you got the dog to just walk up just at your heel, that kept the sheep away. So this was me introduction to sheep.

From now on, much of Ron's time was taken up with the sheep, but he was still responsible for the horses, and used them as much as possible.

When the sheep got to lambing time we started feeding 'em mangels. So it fitted in very well. I put me three horses in the waggon with a big load of mangels, all me grub on the top of the mangels and went to ' sheep, fed 'em and then give 'em what mangels they wanted, louse the waggon out and took the horses home. Now the next morning, I used to go with a pony and cart, feed the sheep, then go to the waggon and take the wuzzels off the waggon and then go home again, and I went with the pony and cart 'til I emptied the waggon. Then the next day I went with a full waggon load, left the waggon there and put me horses in the empty waggon and took the empty waggon home.

Well, one of these days I'd left the waggon in a certain place where I thought was alright and I'd just got home and Boss met me. 'Where have you left the waggon?'

'Well, I've left it in ' parks.'

'Mm,' he says, 'I've just had the Brigadier ring up. Someone's left an unsightly vehicle in the park.'

So I says, 'Oh, I'll move it in ' morning.'

He says, 'You'll move it now!' So I had to drive these horses about a mile and half back to where this waggon was, and put it ' back of a planting where the Brigadier couldn't see it, then *louse out* and go home again. They were the landlords and what they said was law. I mean, you wouldn't think a waggon was unsightly, just stood in the middle of a hundred-acre park? It had to be moved, right there and then.

When lambing time came, the five hundred ewes were divided up between the different farms, and Ron took on lambing those at West Newton, so instead of sleeping at the foreman's house, for the duration he slept downstairs in the farm house, keeping a fire going all night. At the beginning of lambing, the lambs came thick and fast, but once most of them had lambed, those on the other farms which hadn't lambed were brought to West Newton for Ron to finish lambing. Being busy at night and first thing in the morning in the lambing sheds meant that when Norman Caley came round in the mornings, Ron sometimes had not had enough time to do the horses.

Sometimes he'd come round and I used to say, 'I ain't got them 'osses done yet,' and he would help me, turn 'em out while I was feeding 'em. He'd soon muck 'em out. I mean, there was nowt they

couldn't do, nowt they hadn't done, and then it went on from there.
I think I lambed them sheep for about twelve years.

Once Ron was regularly working with the sheep he often took
the Caley children with him when he did his rounds in the pony cart.
As the children got bigger Ron expected them to work with him,
which was a gradual introduction to working life, but when they
were still small, they often fell asleep on the way home from the
fields.

I'd get these kids and we used to put a lot of these Fox's bags in this
cart, and they'd be fast asleep. And when we come back I used to put
the cart gently in the shed and put me pony away and leave 'em. Then
as they woke up, they'd jump out and run home, and they was with
me from being about two to Tim being fourteen. And Norman used
to say, 'If they're cheeky to you, give 'em a good hiding. Don't abuse
'em, but take 'em at back of that stack and while you're hitting 'em, I
shan't have to do. But,' he says, 'their mother'll play hell up with you,
but she doesn't pay you!' And I never touched one of 'em – no!

And we used to have ponies. I broke a hunting horse in, one of
Peter Caley's horses, and I had me a black pony and a grey pony,
and I used to have them little Shetlands on lead reins and we used to
go all over.

Besides using the horses for work, Ron could also use them in his
own time, and once Gil Briggs and his family had moved from
Marton to one of the new council houses at West Newton, Ron
often took Maureen out with the pony and cart.

I used to go round for a ride on Sunday afternoon, and I used to pick
her up with ' pony and trap and we used to go round all these parks,
and Briggs used to say, 'It isn't the lads we know, it's the lads we
don't know,' and as long as they were going about with people they
knew, they weren't worried. We were roaming all around them
parks with ' pony and trap, night after night. Or I used to take two
ponies from Caleys and I used to ride one and she used to ride the
other, and it didn't bother them at all.

Once the evenings were drawing out, the sheep were moved for
the summer onto the fields sown with clover the previous year.

All these farms had white clover seeds, and they used to say the
white clover seeds was the making of these farms. You'd want about

hundred acres of white clover seeds if you'd five hundred ewes, and these white clover seeds were usually undersown under barley, but they could be under oats. All the winter the sheep were in grasses somewhere, and then when it got to be spring and the frosts had gone, you used to bring the sheep onto these white clover seeds. But you used to have to go about six o'clock and walk them off into different grasses, they were walked off every night, and then when the frost had gone, next morning they were walked back onto the seeds.

Unlike the grass fields which were properly fenced, the fields where the seeds were sown as part of the arable rotation had to be fenced securely enough to keep the sheep in.

Quite often there might be twenty acre of seeds that side of a field and ten acre of peas that side. These seeds would be surrounded by corn: there'd be peas and wheat, barley, whatever, all round them and there was this field of seeds in the middle, and your sheep were going to be put in and once you put your sheep in, that was that for the summer. And to save fencing, if they'd slashed hedges in the autumn, in those days they were all cut by hand and they'd probably have five years' shoot on 'em, and as he cut them, he laid the cut in a straight row at the back of him with just enough room to get a waggon between the hedge and what he'd cut. Then we used to go, and they used to fork these thorns onto a waggon. There'd be two men forking and one man loadening. You loadened 'em with a fork, and you had a slash knife there to cut 'em, and you'd maybe set off loadening at one o'clock and you'd still be loadening on the same waggon at five o'clock, and you'd a tremendous load. Then these were taken to wherever and then made a dead fencing between the seeds and the field of corn. They were all put out correctly, laid all one way and then iron pickets put in and tied with baling wire, and that was going to last the summer. And then as soon as we'd finished with the seeds in the following autumn, they would just put a match to them. Because they were dead, they'd burn like billy-o. But that was just to save money, and that was a common practice.

The more usual method of keeping the sheep in was with a temporary fence of wire sheep netting and light stakes. Once these sheep nets were set, the sheep stayed on the clover ley until October, when the fields were ploughed up for the next crop of wheat, the accumulated nitrogen from the clover and the sheep droppings providing nutrients for the wheat. The sheep were then moved onto the new seeds that had been sown earlier that same year, so all the nets needed to be taken up and moved onto the new ley.

> When you was going to sow wheat, you went with your waggons and took so many of these nets up. If it was ten acres, two of you could set the nets in a day, and when you put them out, the man with you measured fifty yards and put a net down. Your stakes were in the bottom of the waggon and the nets were on top. When you'd got rid of your nets, you turned round and drove back, and the man who was with you threw the stakes out roughly every four yards. You'd maybe two waggon loads to get these nets and then you'd set 'em. Then once you'd got 'em set, you took the sheep off the old seeds onto the new seeds. We always *tupped* the sheep on the seeds in the autumn, and then all the nets were wrapped up and taken home. Then they were all taken back at spring and all set again, and the sheep kept in there until the following October, and then that was ploughed up.

Ron's move to being the shepherd and the reduction in the number of farm horses reflects the increasing mechanisation of farming during the 1950s, which also saw a decrease in the amount of labour required. In part this was because fewer men were needed when cultivations were carried out with a tractor, but also some of the work involving human muscle power started to be done by machines. At harvest the introduction of combines, for instance, reduced the amount of manual work, and meant there were fewer days threshing in winter.

> When I first went, there was one combine, and it would do the barley and the laid stuff. They was baggers in them days, and they used to bag 'em up and slide 'em off, and they were all on the floor in sixteen-stone bags. Then you went round and collected 'em up.

Although collecting the sacks of barley with a trailer was still manual work, it was still quicker than cutting all the corn with a binder, stooking the sheaves, stacking and threshing. However, the peas, oats and wheat continued to be cut with the binder, and particularly with a reduced labour force, a considerable number of extra people were needed to get in the harvest. Even with this extra labour, which on one occasion totalled forty men all stooking in one field, with four eight-foot binders running, it was common for the binders to get ahead of the stookers, so often they were not working in the same field where the corn was being cut. Once it started to rain, however, the binders had to stop, giving the stookers time to catch up. All the sheaves of corn which had been left on the ground by the binders had to be stood up to stop the grain sprouting. This was particularly important when it had rained, and it meant that while the binders were unable to go to work again until the standing corn was dry, the stookers had to carry on working in the wet.

> Sometimes we used to have four binders going and there wasn't a man stooking, because they was among peas maybe, and then we used to get these big gangs, and go and stook all up. You was alright while it was fine, but if it rained then what a job, 'cause there's nothing worse than wet stooking. You'd be stooking, and the water would be running down the inside of your legs, and your trousers and your jackets were wet, and you wasn't there long before you was wet through in a morning.

When these big gangs of men were working at harvest time, the task of providing 'luance became a major logistical exercise. With men turning peas, a big gang stooking, and a few others with the binders, there could be 'luance to make for sixty.

> When we were stooking, the little lad used to fetch 'luance round in a morning with ' pony and cart. It was a sandwich and a pint of tea in a morning, and they used to fetch the tea in buckets. You could dip in for your first mug full, but after that the tea was poured into your mug, because not only had you all drunk out of your mugs, you'd stood 'em on the stubble and there was dust on the bottom. And when he come with ' pony and trap he had to come directly into the middle [of the field], not stop at the edge, so everyone had the same distance to walk. We used to sit down for ten minutes, and as soon as somebody stood up, we were off: once they'd got eaten, if any on 'em stood up, they used to say, 'Well, he's keen,' and with that, they were up and going again. They learnt not to stand up!
>
> In the afternoon the same thing occurred again, but we all had a piece of pie, it'd be apple pie, or prune pie, fig pie, date pie, bramble pie, curd cheesecake, ground rice cheesecake, them was all the variations you got over the harvest. And they only baked once a week,

and so by the time the week went round some was getting a bit stale. You imagine baking enough pies for sixty every afternoon, six days a week, and there was maybe three or four lads living in and then she had her own family, and we ate pie for breakfast, pie for dinner, and pie for tea. And there was no electricity, it was all fire oven; ' blazing hot day outside, and they'd this fire going. Very often they had their daughters to help 'em, and some would get a labourer's wife to come in and help 'em, because these foremans' wives were paid so much a lunch. At one time it was sixpence, then it got up to eight or ten pence per man. Well, you get sixty twice a day, six days a week; that added up to some money.

Of all the crops harvested, the peas were the most risky, because if it rained frequently, many of the peas would pop out of the pods as they were repeatedly turned. To avoid this problem the peas started to be put on tripods to dry. The use of tripods became quite common in the 1950s, mostly for drying hay in wet districts, but at West Newton they were used both at hay time and at harvest for the oats and peas. The advantage of using tripods was that the crop was off the ground, and if it did rain, the outside surface of the tripod acted like thatch to keep the inside dry. The way the tripods were used varied in different parts of the country, the size of the tripod and local weather conditions dictating at what stage of maturity the drying crop could be loaded onto the tripod. With carefully built tripods, grass can be safely put on after being dried on the ground for a day; but at West Newton they waited until the hay was nearly half made before being put onto the tripods. Once on the tripods the hay could be safely left until time allowed for it to be buckraked to the stack, though in Holderness each tripod also had two strings over the top, weighed down with bricks, to stop the wind blowing off the tops. Although the name, tripod, should have indicated that they had three legs, the ones Ron used had four legs, the advantage being that they held more hay and there was a bigger air gap in the middle so the hay was less likely to go mouldy. In use, the tripods, which are wired together at the top, have a wire running horizontally between the poles about eighteen inches from the ground, and the hay is loosely forked onto the wires, working round and round, building up the layers. At chest height, often another wire is put over the poles, more hay being put on until the last forkful goes over the very top, making a mini stack about eight feet high.

After the hay was taken to the stack, the tripods were free to be used for the oats and peas. With a large acreage of peas, the men worked in teams to fork the crop onto each tripod, with another team setting up all the tripods ahead of the others.

> We used to do about twelve tripods to the acre of peas, and when we started tripoding, there used to be three of us, and we did nothing but put tripods up. They was coming up at back of us, there'd be maybe three or four lots with three men to each tripod. We'd loads of tripods on trailers, one lad driving the tractor, and ' other lad throwing 'em off, and then we used to have to carry 'em both ways. Then we put 'em all up, so they hadn't to stop.

Once the peas had dried enough they were stacked in the field, the tripods being swept up to the elevator with a buckrake on a tractor.

> We did that with peas for a long while, put a sheet down, put the elevator onto it and every time we stuck our forks in, we used to stick it straight through these sheets. These sheets was all hired, and we used to have endless sheets; you've never seen so many sheets, waggon sheets, stack sheets; and then they'd send a letter and say if you would like to pay x number of pounds, you'd have bought this sheet. I can remember that particular year, a brand-new sheet it was, just laid down and every time you finished a tripod you stuck your fork in. Well you can imagine the amount of tripods that went in a stack, the sheet was like a colander: ' never give it another thought; that's was how it was done.

The change in harvesting methods continued with the arrival of pick-up balers, so instead of sweeping the tripods to the stack, the combine was driven to the tripods and the straw baled behind it. These gradual changes in harvesting were part of the continuing changes in all aspects of farming. For the horses, which had been an integral part of farm life only a few years earlier, their role was steadily diminishing, so by the end of the 1950s even the Caleys had nearly stopped using them.

CHAPTER 11

Horselads and Bosses

WHEN the traditional pattern of hiring horselads was still part of everyday life in the East Riding, the horselads formed a distinct group within society. Although they obviously did have contact with other people, much of their time was spent with other horselads even when they were not at work. Especially during the winter months when it was dark before they finished work, the lads often spent the evenings in the stable. Some of their time was spent feeding and cleaning the horses, but otherwise they were just passing the time. One way of keeping themselves occupied was to mess about with the horses, Ron recalls walking from one end of the stable to the other, balancing on the top of the *skelbases* and the horses' backs. At other times when it was cold, the lads would sometimes lie on the horses' backs, just to keep warm until it was time to go to bed.

> At West Newton we spent whole nights in ' stable, we had a dart board in, and then once they got electricity one of these lads fetched a wireless and had it up on a shelf. And plait; you'd be plaiting their tails, seeing who could plait best, and Norman Caley'd come in and he would plait. And once it got to be spring o' year, it was warm, you'd maybe go on yer bikes, or you'd maybe go down to another farm and look at them, and they'd come to you. Nobody bothered about people roaming in and out. I mean, if Norman Caley was in, he didn't question them, it was an accepted thing; you all went here, there and everywhere.

When the lads went to one of the villages they did not go into the pubs because they did not earn enough money to buy beer, so instead they hung around in groups, sometimes talking; or sometimes some might go window tapping, unless an activity was available in the village hall.

We used to go Sproatley quite a bit, 'cause there was a village hall and they used to play billiards, darts and pontoon. Heck, I could win at pontoon! Oh, you'd win four or five shillings every week, and I did! But it got to be people wouldn't play with me! I couldn't loss, and then all of a sudden I started and I lost every bit of money I'd won, and I couldn't win. So I stopped playing.

Just as at work, the only people in the village hall on these occasions were lads and men, many of whom worked on farms or had worked on farms when they were younger. This largely male environment, both at work and play, and the fact that a horselad had to become a labourer as soon as he married, tended to mean that farm workers delayed marriage. While at work, even if there were maids working in the farm house, association between the sexes was not encouraged. Nonetheless there were many marriages between farm lads and maids, even though relationships between employees on the same farm were often forbidden, so a couple would have to wait until they were working at different establishments, or keep their relationship secret. If they were found out, one of them could be dismissed, and this was invariably the girl. In the 1940s social attitudes were changing, but the physical evidence of the separation of the sexes was still apparent.

As we were going round we used to come across these different farms; very often you went to different farms to borrow implements, or on your bike, and occasionally you saw a window with bars on. And I'd got it into me head that theses bars were to stop lads running away, and when I went to Marton, there was bars there on one particular window, and I said, 'What's them bars on that window for, Foreman?'
'Why,' he said, 'don't you know?'
'No.'
He said, 'How old are yer?'
I says, 'I'm nineteen.'
'And you don't know what them bars are for?'
'No'
'Well,' he said, 'you've seen that three lasses o' mine.'
I said, 'Yes.'
'Well, have you ever thought of whether they're men or women?'
So I says, 'Aye, a few times!'

'Well, that's what them bloody bars are for, stopping yer finding out!' He said, 'That's lasses' room!'

And we did hear tales of these lads, 'cause in years gone by all the maids were hired the same as lads. And some of the farms would have a maid, and if it was Boss's house you might have two, and these lads would get out of their bedroom, specially if there was a roof again' [the window], and they'd climb in these lasses' bedrooms, and he said, 'Anyway, you know now what them bars are for.' . . . I've often thought about the bars!

Despite the distractions, when at work most horselads took an interest in doing their job well. Even if they were not doing it for their own satisfaction, doing a good job would prevent them getting on the wrong side of the waggoner or the foreman. Often the horselads were within sight of the waggoner, but there were many occasions when the lads were working in different fields, or doing a completely different job on their own. By successfully completing such jobs, a lad was proving that he was capable of handling his horses and was a competent and responsible worker. Nonetheless, things did go wrong, often as a result of inexperience, but as long as the lad sorted out the problem he would usually keep quiet about what had happened.

When I was at Flinton, we'd cleaned one of them ash pits out, me and another lad; we shovelled it into a waggon and I'd to take it down from Carr Farm, nearly to Burton Constable Hall, and there was the old Burton Constable brickyard. I'd put one of the big grey horses from Scotland at the farside and a good 'oss at the nearside. But when I got to the brickyard; it wasn't used anymore and hadn't been for a lot of years; but it was still fifteen-, twenty-foot deep and there was saplings growing out ' the bank, and I could not get these horses near enough. 'Cause I'd all this to shovel out over the top into this pit, and I couldn't. So I thought, 'Alright, I'll back 'em up to the edge, let the etch down and shovel it over the back.' So I got 'em swung round, and the thing you do when you're backing a Yorkshire waggon is let the *bearing reins* down and stand on the pole with the bearing reins in your hand and then you say, 'Back,' and soon as they set off to back, the pole goes up in the air and your knees are now level on with their ears, and you're balancing.

I said, 'Back!' and they stuck their feet in, and what did they do? ' Waggon went straight over the top! The pole was right up in the air and I'd to jump for it, and as it was going down for some reason it tippled over, and one of these saplings held it, just on the shelvings. So there was these horses with their back legs down on this bank, their front feet on the top in the field, and the pole was nearly straight over end, and the pole chains, instead of 'em being from the pole up to their shoulders, they were from the pole down to their shoulders, and the horses were nearly being choked, and I thought, 'Now what the hell do I do?' I went round the back and knocked the pin out ' the etch, and it just went whoosh, every bit straight down, and I thought, 'Now then, how the hell do we get out of here?' 'Cause we were only on two wheels, and the one sapling holding me. So I cut a big ash plant and I let 'em stand there a bit. Everything was nice and quiet; there was nobody around, and I just had me strings in me hand and I got ' ash plant and I shouted, 'Go on,' and hit 'em both at the same time, and by hell, they jumped out ' their skins. They jumped, and the waggon righted itself and we was on the top. So I put the etch up and I thought, 'Well, I haven't no shoveling to do.'

I went back, and it was dinnertime by this time, so Foreman says, 'How ' you going on?'

'Why, I've another load to get,' I says, 'I think I'll change horses.'

'Aye,' he says, 'that'll be a good idea.' So I put another pair in and loadened up again, 'cause I thought if I couldn't get up to the side the first time, I aren't going to get nowhere near the second time, and of course, this pair was different again. We walked straight up ' the side and I shovelled it over the side, and come home again and nobody knew. I never did tell anybody the episode I'd had.

Although Ron was the one hoping no-one would find out what had happened on that occasion, the boot was on the other foot in the case of a lad who came to learn farming as a farm pupil.

We had a lad, he was supposed to be a farm pupil, but he lived in, and Foreman says to him, 'Put a 'oss in that cart, and go and lead them tatie tops off that field!' When we finished picking the taties, we harrowed the field to pick what was left, and all the tatie tops were left in rows. So all you had to do then was go with your cart or waggon and fork 'em in. So he said to the lad, 'Go with that cart.' It was a rubber-tyred cart; 'Go and tip 'em in that dry pond!' So off he went; and ' dinnertime I come up, and he says, 'Will you come and help me? My cart's in the bottom of that pond, upside down.'

I says, 'Where's your 'oss?'

'I got it out.'

'Well, go for our dinners,' I said, 'don't you say anything!' So when you would go in for your dinners, Foreman would say, 'How did you go on?' and you'd just tell him.

'Are you alright then, Michael?'

'Yes,' he says. So after dinner, when we come out, I went, and there was the cart completely upside down. I said, 'How the hell did you get the 'oss out of there?'

'I don't know, but I got him out,' and the horse was in ' stable, no bother.

So I says, 'Well, put some traces on it!' So we went down and pulled the cart onto its wheels, and then we hung to the cart and we pulled it out ' the pond and put the horse back in, and I said, 'There, you're alright. Next time you do it, you want to get a sleeper and put it at the edges of these ponds so that you back up to the sleeper and tip over the top.'

Learning how to do the job was not the only thing this farm pupil found out the hard way. The idea behind being a farm pupil was to learn how to become a farmer and while many pupils would expect to live in the farmer's house, anyone living at the foreman's house had to learn how to fit in with his household. This meant they would be kept in line by the waggoner, who by this time was Ron.

This lad hadn't been used to our way of living and if there was a pie, you cut you own piece, and if it was burnt at one side he would turn the plate round and cut a piece that wasn't burnt. Well somebody had to eat this burnt piece; if it wasn't today it'd be tomorrow. So Foreman says to me, 'Have you watched that lad?'

'Aye.'

'You aren't going to let him do that, are you?' This was teatime.

I said, 'We'll stop him. Where is your biggest spoon you've got?' So he give me one of these big spoons and I put it down again' me plate, and they used to cut a bread loaf up, and on the top was one crust and somewhere down the heap was another crust. But when he come to get his piece he pulled the next piece from under the crust and left the crust, and by, didn't I hit him with this spoon, right on ' back of his hand, and by, he jumped. I just said, 'Take it as it comes!' And nobody said a word, but by hell, he didn't do it again. I bet his hand hurt. Never mind, it cured him.

Keeping lads in line by physical means was by no means uncommon, though once the waggoner had established himself and had the respect of the other lads he would lead mostly by example. Relations between the lads could then be more friendly, but for the waggoner there was always a fine line between being too friendly with the lads one day, and needing to discipline them the next. This conflict was clearly illustrated when Ron was at West Newton and had a small problem with one of the lads. Although it started out as a joke, in the end the waggoner's role as an authority figure had to be preserved.

> He was a lovely lad was this, Billy Tate they called him, and he was best lad I've ever had, but he'd done something, and I said, 'If you do that again, Billy, I'll kick your arse 'til you can't stand.'
>
> So he said, 'You'll have to catch me first,' and by that, he'd gone down ' stable. I used to wear big scotch boots at that time and the first inch of the sole had a plate on and they was all hob nailed; and Billy was running, and I run after him swinging my leg. I didn't intend to kick him, but he stopped, and by, I kicked him and he rolled down in pain, rolling on stable floor, and I said, 'I told you I'd kick your arse if you didn't behave yourself.' But by hell, I was sorry; I never intended kicking him.

Essentially, the waggoner's job was to set an example and keep order, but in reality his role was more complicated. Although he was there to carry out the orders of the foreman and the boss, and maintain a high quality of work and discipline among the lads, in the respect of feeding the horses it was he who often set an example in deliberately disobeying instructions. In earlier times there was a widespread practice of feeding horses drugs. This was done by the individual lads in secret, in order to increase the horses' appetite so they would appear to be well looked after, or to give their coat a sheen. Many of these drugs were poisons, so had to be administered carefully. However, others were addictive, so while they might serve the interest of one particular lad, when he left at Martinmas his horses were left to go 'cold turkey' over the following weeks. Although it was illegal to feed drugs, horselads would cycle miles to different chemists to obtain the individual constituents of these cocktails, the recipes being passed down the generations.

On the places where Ron worked, however, there was no use of drugs, but the illicit feeding of wheat was commonplace. Like the giving of drugs, this was a dangerous practice, as the wheat would swell up inside the horse and give him colic. So feeding wheat had to be done carefully, making sure the horse had had a drink before being fed the wheat. Despite the danger, as long as the horses were not made ill, some foremen probably turned a blind eye to the pilfering of wheat and other feeds such as linseed cake which the lads fed to the horses. Nevertheless it was not good for the horses, which have evolved to eat large quantities of roughage, low in feed value. Even oats, which were necessary to provide enough energy for a horse in full work, had to be fed in small quantities, little and often, for the horses' digestive system to cope with it, which is why Ron was not entrusted to feed any horses when he was fourth lad. But at Carr Farm, despite there being no limit to how much oats they were allowed to feed, a little while after Martinmas the waggoner started giving them wheat.

> When we'd been there about a week the boss comes down ' the stable and he says, 'You won't feed any of these horses wheat, because if I catch you feeding them wheat we'll have you all at Sproatley Police station!' So he walked out, and when he went out, Wag says, 'Now we can start.' I wasn't feeding any; Wag and Thoddy was feeding, and after a day or two Wag come round and said, 'Don't give that one no more, and don't give <u>that</u> one no more,' and their legs were coming up and they wouldn't stand it.

When changing the feed given to a horse it is standard practice to change it gradually, but it was even more important when the horses were fed wheat. The feeding of wheat started with the new moon, only a very small amount being given at first, increasing gradually over the following fortnight.

> I think the moon was just a calendar. Whether you could've done the same looking at the calendar I don't know. But the thing was you started them with an egg cup, and by ' time you got to ' full moon they were eating a pan full. They'd had a drink first, and then you fed this wheat, whole wheat. Oh, you could see where they'd been. I mean, when you think these horses were eating a stone and a half of oats and then a tin full of wheat, and when I worked at Marton and I was on me own, they were getting wheat in the morning and a tin full of barley at night and as many oats as they would eat, and again they were ploughing all day long. And once we finished ploughing, as soon as we had a good frost we started quarting and turning

155

furrows back and so them horses, ' only time they were stood was at Saturday afternoon and Sunday. All winter through they were completely going.

Even though the horses were working very hard and needed a lot of food to do the work, this heavy feeding, especially with wheat and barley, was dangerous, and led to the horse's legs filling with fluid, and cases of colic which could and did kill horses. All the horselads knew this, and knew they could face prosecution for feeding wheat, but it was still a common practice.

In the feeding of horses, maintaining discipline and maintaining a good standard of work, the waggoner's role was crucial. Besides being capable of carrying corn on threshing days and having the ability to keep the lads working effectively, a good waggoner also had to be able to think for himself. These abilities were essential if a waggoner intended to become a foreman after marriage. But the intelligence needed to become a good foreman or waggoner was by no means universal; in the East Riding there were very few jobs outside farming, so farmers employed men covering the whole spectrum of abilities. With routine jobs done day after day most men could be usefully employed, and when much of the work was done in groups, there was often someone who had a bit more about them who could keep an eye on the others. But that was not always the case.

A lot of them fellers weren't very bright to start with. There was a lad when I was at Flinton, he was as thick as they come, but you give him a pair of 'osses, couldn't he drive them! When he talked to them, you'd think he was a hundred years old; but if you had to tell him to do something, you had make sure you went and see he was doing it; it was no use just telling him to go, 'cause when he got there he didn't know what the hell he was doing. I allus remember, we were in some fields and they were four hundred yards long, and Wag says to him, 'While we're filling these drills up, Tommy, go and measure!'

Tommy says 'Aye,' and he's striding off, and when he got to yon end he stopped. And Wag said, 'The bugger's forgotten, look!' He come back, and Wag said, 'How many yards was it then?'

'I forgot when I got to yon end!' But the bloody feller hadn't the sense to measure back again!

He said, 'You daft bugger,' and so then I had to go and measure it. Well, he was thick!

Because of the importance of the waggoner, farmers would try and keep a good one from one year to the next; though as a rule most horselads moved every year. If a farm was known as a good farm and had good horses, there was some status in working there, but that was not always enough to persuade a waggoner to stay put.

> Nuttles Farm, between Preston and Lelley, was supposed to be one of the best farms round here, and they'd always had good horses. George Gibson said they were about ten or twelve horses all under five, and they had always had a good waggoner. Then they got this waggoner and he wouldn't stop, and they tried to persuade him. Then they got to Hull and tried to persuade him, and he wouldn't come back, and eventually they hadn't much choice and they were left with another man, but he was no good whatsoever, no good in their eyes. He was a bit thick; and when they were drilling roadside fields, he couldn't drill straight enough for 'em, so he was only let loose on the insides. He was a real let down for them.

Just as the farmers wanted to keep up standards, the horselads, most especially the waggoner, also expected things to be done the right way and to be treated with respect. Given the pride they took in their work, any belittling of their efforts, whether intentional or not, could be seen as undermining their position and status.

> At Pasture House they'd never had a tractor, everything was done with horses. And they went to the wagg'ner, and they said, 'Would you like to go to the ploughing match?'
>
> He said, 'Let the buggers go who come,' and they'd fetched a tractor in to do some ploughing: he wouldn't go 'cause they'd fetched a tractor in.

Even though the bosses had little to do with the horselads on a daily basis, the decisions they made could have a big influence on the lads' lives. Especially during the depression, farmers tried to minimise their outgoings in any way they could, including reducing the amount of money the foreman was given to feed the lads, and avoiding any extra mouths to feed.

> Some of them fellers, there was two or three brothers all hired together on the same place at times. They'd be there maybe for a year and then they'd go their different ways, then go together again. And [on] some of them farms, Foreman's sons would maybe be

hired at home that year and then they'd leave. But they couldn't live at home. If they wouldn't work there, they couldn't sleep there.

In the 1930s it was not just the farmers who needed to avoid any extra expenses, but the parents of the horselads, many of them labourers or foremen themselves, also had to count every penny.

Summertime, and there was no 'osses to do, people used to go home at weekends, but they couldn't live there. They used to go home but they'd go back to their place for their dinners and go back for their teas, 'cause that was all a part of their fare, and I've heard 'em on about biking from Sproatley to Little Humber, that's up Paull way, and they come home and go back for their dinners. There was no money, no money at all.

Besides the hiring of the lads and paying them at the end of the year, typically the bosses let the foremen deal with the horselads. Especially on the large farms all the orders were given by the foreman, some farmers only being involved with buying and selling, and deciding which crops would be sown where. On the farms where Ron worked, however, the Caleys were all actively involved with the daily farm work; in fact, they probably would not have been farming at all if they had left all the work to others during the depression years. Most of the orders and the organisation of the day's activities, however, were still left to the foreman, though having done all the jobs themselves, the Caleys would intervene when things were not going right.

The boss didn't tell you anything; I say didn't tell you anything, if it was going wrong in the field he would tell you! But at one time, they used to say he would see you going wrong and go home and fetch Foreman to come back to put you right, and then all the falling out was with Foreman.

Although during the year the bosses were definitely in charge, albeit mostly at a distance from the horselads, at Martinmas the two sides met on much more equal terms. When a lad was asked to stop again, this was the time for him to negotiate the best wage he could and to iron out any other problems, but if he had declined an offer to stay for another year, the boss might still hope to persuade him to change his mind during the course of the conversation on Martinmas day. For those lads who were inclined to stay there was no point in being too willing to accept the first offer, so during this discussion there was still everything to play for. The way the boss conducted the discussion could also affect the lad's decision, as when Ron had finished his second year working for John Caley.

When it come to Mart'mas, he says, 'How much do I owe you then, Ron?' So I said, 'Well, I don't think I've had any subs,' and he had it all written down. He was at that side of ' table and you was at this side of ' table, and there was all the stamp cards there, and all the money there, and he says, 'How old are you? I've been stamping your card as though you were nineteen.'

'Well, that's no good,' I says, 'I wasn't nineteen while October.' My stamp was maybe tuppence more than it should have been.

He says, 'What are we going to do about it?'

So I said, 'What do you mean?'

'Well,' he said, 'I've been putting the wrong stamp on.'

'Aye, whose fault was that?'

So he says, 'Well, you'll have to pay for that.'

'I haven't stamped me card,' I said, 'I haven't touched me card. If you've put it wrong, then it's your mistake,' and it come to be something like four shillings and tuppence.

'Well,' he says, 'what are we going to do?'

'Well, I aren't going to do anything, what are you going to do?'

'Well,' he says, 'shall we split the difference?'

'Nowt to do with me,' I said, 'you're stamping the card.' And he was going to try and make me pay the four and tuppence more, and we was arguing 'bout this bloody money, and I says, 'It's up to you to put it right,' and eventually, it took him a long while, 'Oh well, I suppose I'll have to do, then,' And then what do you think he says?

'You're going to stop again, aren't you?'

And I thought, 'You daft bugger. You go and argue over four and tuppence, then say are you going to stop again!'

Once Ron had turned down this offer, John Caley then asked if Ron wanted to go to any of his brothers' farms, because just as the lads would tell each other about the different farmers they had worked for, the Caley brothers helped each other in trying to keep good workers. Ron having declined these offers as well, John Caley then asked him if he wanted to go horse-rulley driving in Hull, or to go and work in the shipyard at Paull, using his influence in the community to secure Ron a good job elsewhere. But by that time Ron had made up his mind not to take up any of these offers, and instead take his chances at the hiring fair the following week. When he got there, of course, probably in a calmer frame of mind, John Caley took him for a drink and Ron agreed to go to his brother Norman. Even with Norman Caley there was some animated discussion at Martinmas, but, as on this occasion, the outcome was different.

> When I wouldn't stop with Norman, we'd be arguing, and the stamp card would be going from that side of the table to this side, I'd be pulling it back again, we'd argue and he'd get it back again, and I was just going out, and he says, 'Come here! Here's five pound for you, you've been a good lad.' He just said that as I was going through ' front door. He said, 'Now are you going to stop with me again?'
>
> I said, 'Now you're talking,' and I walked back and sat down and stopped with him again. Now if John had had a bit more about him, instead of saying, 'You owe me four and tuppence; here's five pound for being a good lad,' I would have stopped. But on principle, if he's going to bloody argue about four and tuppence. Like, I couldn't believe it then!

The discussion on Martinmas day was strictly between the farmer and the lad. Although it might have been the foreman who asked the lads if they were going to stop again in the weeks running up to Martinmas, the contract the lad made was between him and the farmer, and even the farmer's wife was not expected to be involved, although one year Mrs Caley was in the room with Norman and Ron.

> I remember once he paid me, and she was sat in the room and she kept saying, 'Will you two shut up?' and, 'Get him paid, Norman!' I mean, he didn't take a happ'oth of notice of her; he'd be on about something and I'd be on about something, and then we'd argue again, and she was sick as hell of listening to it. Well, she never ought to have been in there.

Ron's indignation at having Mrs Caley overhearing the conversation shows how unusual it was, and also that a lad had a very clear idea of the relationship between himself and the boss. Nonetheless, on another occasion shortly before Martinmas, Mrs Caley's involvement proved to be to Ron's advantage.

> It got to be nearly Mart'mas and she said, 'Are you going to stop again?'

'He's never asked me.'

'Oh, he will do.'

'Aye, well I aren't going to stop.'

She said, 'Whatever for?'

'Well,' I said, 'you want to sleep on that bed. It's flock, and as long as nobody touches it, you can lay in the same hole that you was laid in the night before. But if they shake it up, it's like sleeping on cobbles.'

'Oh, how bad a bed is it?'

'You want to come and have a look.'

'Come on then,' she says, 'there's nobody in.' So we goes upstairs, ' have a look at this bed. I says, 'You have a lay on there and you tell me what you think.'

'That's no good is it?'

'No good at all.'

'Well, have you told him?'

'No.'

'Well,' she said, 'I want a new bed. You tell him you aren't going to stop because you want a new bed. He'll come and tell me, and I'll tell him he can have ours. You can have the feather bed, and we'll have a new one and he won't be no wiser.' And that's what happened. I get the feather bed; she gets a new bed.

Although on this occasion Mrs Caley acted as an intermediary between the boss and the lad, and in the day to day work on the farm the foreman performed a similar role, the individual characters of the bosses and the conditions they found themselves in meant that they all had different ways of working and dealing with the men. Those who had gone through the years of the depression knew that it was only by hard work that the work would get done, a common phrase being, 'Come here me bonny lads, let's be having you. Here it is Monday, ' day after tomorrow's Wednesday and bugger all done yet!'

Whereas some bosses were quite friendly, others such as John Caley would push hard to get the work done. So it was a surprise to Ron to hear what he had to say just before Martinmas.

He said, 'If everybody ploughed like you Ron; you can do as much work with a pair of 'osses in a day as any man I've ever had.'

I thought, 'Hell', 'cause Norman was a king to what John was. Oh, John was mustard. And I used to have little bloody jibes at him. Once I said, 'A bloody good job there isn't a Charlie Wiles's?'

Charlie Wiles was the knacker man from Woodmansea who came twice a week to take away any dead livestock.

> He said, 'What do you mean?'
> 'Well,' I says, 'Charlie Wiles's for men!'
> He said, 'Why?'
> 'Well,' I says, 'you'd have these buggers there now; all these buggers, they'd all be shot.'
> 'Ron,' he said, 'whatever gives you that impression?'
> 'Well,' I said, 'if I was in your place, I'd have the buggers shot as well!'

These interactions show how the ethos of hard work was not just confined to the bosses, and also that the horselads were not especially deferential, though Ron was probably unusual in his bold attitude towards his bosses. Nonetheless, the bosses were quite capable of giving as good as they got, and did not mind as long as the work was being done.

> I used to work a lot harder in winter than ever I did in the summer. I weighed ten stone two in the summer and nine stone two in the winter. By hell it was hard. I mean it suited me, I loved it. Norman used to say, 'If you don't go steady, silly bugger, you'll kill yourself,' and he often used to say, 'Whe're you bloody going now? Go bloody steady!'
> And I used get onto him 'cause he didn't work. I once took a ladder away and he says, 'How am I going to get down?'
> I says, 'Jump! There's a heap of straw down there.'
> 'Aye,' he says, 'I might break my legs.'
> I said, 'Well, I jumped down. I didn't break my legs.'
> 'Well, I weigh more than you.'
> 'So your legs are made in relation to your body.' I said, 'You get your bloody self down!'
> 'Get me a bloody ladder!' he says.
> I says, 'You'll be up there for all the afternoon!' Hell, we had some bloody do's. It's a wonder they never sacked me. Hell, I did used to chew the buggers up.

Another occasion when Ron challenged Norman was when loading sheaves onto trailers one harvest. The trailers were pulled from stook to stook by a pair of horses, and held twice as many stooks as a waggon, but Norman Caley still did not think they were making big enough loads.

> He used to fork all harvest, and one day he comes along, I was loadening, and he says, 'I could put more bloody sheaves on a waggon than you buggers are getting on there!'
>
> 'Oh,' I says, 'I've heard your tales before!' ' Never said no more. [We] carried on, and we'd just about got harvest over with the last field, and I used get up in a morning and get these horses up, and I was putting these horses in a waggon, and Foreman said, 'What are you doing?'
>
> I said, 'It isn't what I'm going to do, it's what our boss is going to do.'
>
> 'Alright,' he said, and off I went, and when I got down to ' field, Boss said, 'Now silly bugger; now what are you doing?'
>
> I says, 'It i'n't what I'm going to do, it's what you're going to do!'
>
> 'And what am I going to do, then?'
>
> I says, 'You're going to show me how to loaden.'
>
> 'Oh, is that what you want?' He says, ' louse 'em out and get 'em in that trailer!'

As the day went on, once the empty trailers were not coming back into the field as quickly as they were going out full, Norman Caley knew they were gaining up on the stacking gang.

> He says, 'Do you think we've got that stack bunged up yet?' He says, 'Shove them 'osses back in that waggon, and when we get high I don't want you shouting, I can't reach, I can't reach.'
>
> So I run down these stooks, and every twenty-five stooks I put a sheaf on the top, and I run down a hundred stooks. Now a waggon load was something between forty and fifty, that was considered quite a big waggon load, and he got in this waggon and I didn't half belt 'em up to him.
>
> He was loadening, and he was loadening . . . kept loadening. Got fifty stooks on, got sixty on, got seventy on. Then the shelvings of this waggon started to come down, 'cause there isn't many inches between the shelving and the waggon wheel, and it was acting as a brake, and these 'osses were having to stick their feet in and jump.
>
> And I says, 'You're going to smash this waggon.'
>
> 'A likely story,' he says, 'you're buggered!'
>
> I says, 'I aren't!'

'And I aren't loadened either,' he says, 'keep em coming!'

So I says, 'These 'osses can't pull it.'

He says, 'Hit 'em then, move 'em! I've nothing near a load up here.'

I says, 'Well, the shelvings are on the wheel like a brake.'

'Alright,' he said, 'what time is it?'

'It's about dinnertime.'

Alright,' he said, 'louse 'em out,' so we had to take these 'osses back to West Newton, and he says, 'Fetch a drawbar. I'll see Ron with his Massey tractor.'

So when we gets back, he says, 'Take the pole out, shove that drawbar on, and hang to this Ron.'

Ron Hunter had been with them a lot of years, and Ron used to keep him right. He used to say, 'What the hell are you doing, *maister*?' He says, 'You'll have it over, and you lot won't want to pick it up, somebody else'll have to pick it up. I should stop playing silly buggers and leave it at that!'

'Now,' he said, 'help me to get back up on this load,' so we throw a rope up over and he climbs back up and he says, 'Right, now let's have 'em,' and we got the hundred stooks on.

'There, will that do yer? And after dinner he brought his camera back, and he jumped on the horses' backs and [I] took ' photograph and that was the end of that; he'd done what he'd said, and they took it home with a tractor.

As much as these incidents show the relationship between the bosses and the lads, it also reflects the individual qualities of the different characters. Whereas Ron was more than happy to stand up for himself, and enjoyed his time working on all the farms, he recalled a conversation with a man who had been at Carr Farm just before the war, who had a different experience, showing how the teller of the tale is just as important as the person they are talking about.

'You could never do enough for John, never,' and he was telling me all about how nasty John was. I had two year wi' John, and he was funny, but Porters were funnier than John had ever been. And then he says he went to George [Caley] and he says that's a different kettle of fish; George could get all the work out of you, but a different approach to John. George would come and talk to yer, and kid you along a bit, different feller again.

And swear: every other word George was swearing. Oh hell, aye, every other word, swearing. Police pulled him up once for swearing on the telephone, or something or other, and he says, 'Who's fucking swearing?'

Hell, he was ever such a nice feller was George.

CHAPTER 12

Foremen and Labourers

FOR the overwhelming majority of horsemen in East Yorkshire, their time spent working with horses was finite. As soon as they got married, whether in their early twenties or into their thirties, they would have to work as labourers or possibly get a job as a foreman. Many of the weddings took place around Martinmas, as only then was a lad freed from his yearly contract and able to live as a married man. For the few that did marry during the year, they had to continue living as if they were single until the year ended.

> At one time of day no matter what happened, if you'd to get married in the middle of the year like some on 'em had, you didn't leave. You got married, but you come back to your place and you slept there, and this lass, if she was a maid somewhere, they wouldn't let her out either; she had to stay. Even Waggoner at Pasture House when I was there, he got married during the year and he had to go and ask permission. She didn't live far away, so he came back every morning and did 'osses before breakfast, and he lived in the farm house, but he used to go home every night, and he did that 'til Mart'mas.

There were a few exceptions to the rule of only employing single horselads, such as on some large estates which sometimes employed a married waggoner on the home farm. At West Newton there had also been a married waggoner during the time when the Caley brothers were still young men living at home and doing the foreman's job themselves. So in this case, besides his usual duties, the waggoner also acted as hind, being responsible for feeding the lads living in the hind house.

Some of the married labourers lived in cottages in the villages, but others lived in cottages tied to the farm. Especially before the war when there was no overtime to be had, the labourers often fed the bullocks at the weekend in return for their accommodation.

> They did the bullocks at ' weekend for their house rent. They got their taties and a pint o' milk a day, and then they fed so many bullocks. Some were doing it every other weekend, or some would be doing that yard o' bullocks, and some would be doing <u>that</u> yard o' bullocks; and they did it every weekend, Saturday night, Sunday morning and Sunday night. And then you'd find some places, they got their house, and some on 'em paid 'em overtime for doing the bullocks at ' weekend.

For the younger labourers who had enjoyed working horses before marriage, there was still the possibility of working them when there were not enough horselads to do the work, typically when ploughing, or when spreading artificial fertiliser while the lads were harrowing. Whether a labourer was expected to work with horses depended upon his terms of employment, some of the younger ones specifically being contracted to work with horses when necessary. Although there were some labourers who enjoyed this work, there were plenty of others who were only too pleased not to be working the horses anymore. In the East Riding, particularly before the Second World War, there were very few jobs outside farming, so unless a lad was prepared to leave the district there was little alternative to farm work, and most of the jobs on the farms were with horses. Although most became competent, it did not necessarily mean they actually liked working horses.

> Very often these waggoner fellers, if they got married when they were about twenty-one or twenty-two, they'd only had a year or two as waggoner, and then they were labouring. A lot of them were hired to go with horses when required, so there was no *jibbing*, and I've heard him say, 'You'll take them horses, will yer?'
> 'No, I haven't come to go with horses, Foreman,' and he hadn't, but <u>that</u> one had; he'd been hired to go with 'osses as required. But these others, they were maybe forty; they hadn't been with 'em for ten years, not really with 'em; they didn't want 'em. They didn't want 'em when they was with 'em; it was just a job.

Even the labourers who did work the horses were not responsible for harnessing or feeding them, but they were supposed to take their harness off before leaving work at the end of the day. But in their haste to go home, there were some who just left their charges at the stable door, and had to be reprimanded by the foreman.

> There was one or two young lads who lived at home, they were 'oss labourers, and they went with 'em, but they were all geared up for them in the morning when they come. But at night everyone had to strip their own horses and if anyone didn't strip 'em at night, some of them married labourers would just whip 'em in, and they'd gone; you used to tell Foreman, and by hell, next morning he got his name for nowt, in front of everybody.

Most of the labourers' work, however, did not involve horses. In winter, after forking straw from the straw stacks, bedding up and feeding bullocks, much of their work was trimming or laying hedges and cleaning out dykes, except on threshing days when they were forking sheaves from the corn stack onto the threshing machine or stacking straw. During the summer there was hoeing, shearing, hay-making and harvest, the work changing from week to week, so the labourer had to possess a variety of different skills, and were usually knowledgeable about many different aspects of farming.

A good example of this is Sammy Hoe, who had started worked as a back door lad at West Newton, doing odd jobs around the farm and the house, including cleaning shoes and pumping water for the family's baths. He then went on to be the bullocky lad, shepherd and had been a horselad both in Holderness and on the Wolds, and had travelled a stallion for the Caley family round the district during the covering season. Because of his experience, he was often able to sort out a problem, such as the occasion when Ron had to fetch a boar back to West Newton from one of the other farms.

> We'd got up to about a hundred and forty sows, and some we had black boars in and some we had white boars. So I went to fetch this boar home from Low Fossam. I went in and got it out of ' fold yard, walked it through the farmyard gate and down the road and he went lovely, just touched it with a stick. We got to the far end, right to the last gate out of the farm and it would not go. It turned round, and it squealed and it squealed, and eventually we get back into ' fold yard.
>
> We start again, and get right up the road, just me and the pig; we'd gone maybe half a mile up this road no bother, right up to this gate and it started again. It squealed, squealed, and run back, and I spent nearly all afternoon; I never did get this pig through this gate.

So I walked home, swearing. And Sam was nearly always there. 'That bloody pig, Sam. I get it far as Low Fossam gate,' I says, 'and it squeals and runs back.'

He says, 'Take some string and tie its ears back.' So next day I goes down, goes to ' black pig and rubs, you could rub their backs and make 'em lay down, then tie string through his ears and right back up, 'cause a black pig, his ears are straight down. I got him out, went to this gate, straight through it, and straight home! Sam was there. 'Now then,' he says, 'you've got your pig home!'

Oh, he knew all sorts of things did Sam, nowt he didn't know, and he didn't always tell you, never a *clever feller* like, but if you asked him, he would tell you. When a 'oss was wrong they used to go and get Sam to look at it, and if owt was wrong, a bullock or anything.

No matter how skilled and knowledgeable the labourers might be, the crucial person in the effective running of the farm was the foreman, who did not earn much more than the labourers even though he had additional responsibilities.

These foreman fellers you lived with, they hadn't but about five shillings a week more than a labourer, but they had a cow, and the cow was part of their wages. They had to milk it before breakfast and after tea and the milk was theirs. So if it give a bucketful, you'd a bucketful of milk; if it give half a bucketful, you'd half a bucket; up to him to feed it right to get the milk. And also he killed two forty-stone pigs, and you never kill a pig unless there's an 'r' in the month. So one pig was killed in November and the next pig would be killed in January or February; you wanted good frosty weather so that the meat would keep. And then he also had a dozen hens and the cockerel when he went, and he could keep up to fifty cockerels while Christmas. He could kill these and sell 'em at Christmas for some Christmas money, but after Christmas he had to be down to just the twelve hens.

Having a good foreman who knew his work, who could manage the men and organise the farm work, not only benefited the farmer but also made life more pleasant for the labourers and the horselads. Although the labourers were free to move every year, if they had a good foreman they often stayed longer on one farm.

Up 'til George Gibson going to Flinton, John Caley never kept anybody. Every labourer left every year, all the lads left, and ' foremans left about every two years; he never had nobody. Then he got George Gibson, who'd only been across the road at Richardson's, and once he got to Carr Farm, for several years all the labourers stopped.

Besides being well respected by the men, George Gibson also liked to show the lads how to do a job, and in doing so, demonstrate his own capabilities. Being a keen horseman, he competed at ploughing matches, and when ploughing on the farm enjoyed showing the lads his skill with a plough and a pair of horses. This was despite having to take any two horses that were left in the stable, often two that did not usually work together, after the waggoner, third lad and fourth lad had taken their usual pairs.

> If you was a good foreman, and hell he was, he could plough; and these 'osses seemed to be all over and he was just chewing his twist and spitting and carrying on. And he would sometimes say, 'I'm a premium rigg man.' That's what they used to say, and he went one day and he set a rigg; you open them out and then you close them up, and go five times around; and when he got five times around he says, 'Now then, which way are they ploughed?' And you couldn't tell whether they was all turned to the left or all turned to the right, they were so identical. Then he got a swingletree and he put it on, and the rigg was no higher than the rest, and he said, 'That's a premium rigg!' That takes some doing, and it was no effort to him.

Besides showing the lads how to do a job, he was also keen for them develop their own capabilities by making them do a job themselves. One such instance occurred when they were leading waggon loads of mangels into the fold yards with three horses at the beginning of the cold spell in 1947.

> The first time we went, Foreman said, 'Take your first 'oss off when you go in among ' bullocks,' and we took him off and tied him up outside and drove in with two horses, threw all your wuzzels on to ' floor for the bullocks to scalp, and put him back on when we got outside. But after we'd been a few weeks, we used to go in with all the three, and go round the pillars, in the sheds and out the sheds and round the tum'rils, and we got to be real heroes, or we thought we was.
>
> And I remember one day coming with a big load of hay; we'd

been to the haystack; we'd gone with three horses and we'd been fast in a few drifts and had to louse out and pull the waggons backwards. You used to take 'em out and hang to the airbreeds at the back, pull the waggons back, and then stick the 'osses back in. And I come back with one load, and Thod Lad was going to run out and get hold of me fost 'oss'es head, and Foreman said, 'Leave him alone!' and you had to drive 'em yourself. If you was going to do a job, you did it, you didn't have nobody leading 'em and messing about. He made you do it, did that Foreman.

The other way the lads learned was by listening to the stories of the older men, though some, including those told by Ted Simpson at Carr Farm were as much fiction as fact.

This little feller who'd been Charlie Buck's lad in 1900 had also been hired and been round parts of Holderness, and he'd now come back to where he started as a thirteen-year-old lad. And he was very bad on his feet, and his head used to shake, and his hands used to shake. He was bullocky/chicken man and he used come into ' stable and he'd sit on a bale; we'd be all there in a morning, maybe twelve on us and he'd start, 'When I was Waggoner at such and such, Foreman,' and his head used to shake, and we used to say, 'The more Ted's head shakes, the bigger the tale he tells,' and he was sat there and he had an audience of all these labourers, and they'd heard his stories many times before.

And he said, "When I was Waggoner at such and such a place I had a pair o' yellow bay horses," and Foreman said, "We'll start leading, Ted."

"Alright, Foreman," and he starts to loaden . . . and he was loadening. It was a big long field, and Foreman says, "You're getting a big load on, Ted."

"Aye, well," he said, "I've got nothing like a load on yet, Foreman."

"Well," he said, "these horses can hardly pull it."

"Well, you'd better get some more!" So Foreman goes home, and they put four horses on this waggin, and Ted's still loadening, and he carries on and they fetched two more horses, so he had six horses on, and he's still loadening, and he said, "I haven't got a load yet, Foreman," and they fetched two more horses; put eight on, and Foreman says, "They aren't going to pull it, Ted."

"Aye," he said. "well, I haven't got a load on yet."

"Well, the only thing we can do is to louse out and we'll fetch the threshing machine and we'll thresh it where it stands!"

Well a day's threshing is twenty-four loads, and Ted had put all this on one waggon, and eight horses couldn't move it!

And he says, 'It's as true as I'm sat here, Foreman.'

'Aye, well,' Foreman would say, 'that's enough of your *lees* now, Ted!'

Despite bringing Ted's storytelling to a close, George Gibson, like many in rural communities, also relished telling stories from when he was younger.

If you could get Foreman going you knew full well you'd have a good 'luance time; and them labourers used to know an' all. They'd say, 'And then what, Foreman?' and get him going. They'd heard these tales time and time again, but while they were sitting, they weren't working. Then all of a sudden he'd look at his watch.

'By hell!' He used to want to be off, like, and jump up. Oh aye, George Gibson could tell you the tale.

The labourers and lads might be able to pull the wool over George Gibson's eyes, at least for a short time while he related one of his stories, but he was nobody's fool. Anyone trying to pull a fast one would soon find they had chosen the wrong adversary, as these two incidents with potato dealers show.

When we were green topping we dealt with three firms, and this particular firm, there was always something wrong, and whenever you talked to anybody who dealt with this firm, this feller always had some excuse. The other firms used to come in and they'd let anybody drive their lorries, they would loaden 'em and we would have a [hicking] stick and hick 'em up in hundred weights to 'em. We would put five ton on, that is what them lorries would hold. Well, this one he wouldn't come in; he said their boss says they hadn't to go in any field. So Foreman says, 'Well, alright, we'll put 'em on ' waggon.' So we pulled a waggon down and when we got these taties, we carted 'em out and transferred 'em onto this waggon. Well three ton, or five ton on a waggon is a tremendous load, and up come the lorry, and Foreman says, 'They're on that waggon,' and he pulled up to this waggon. Now, down your *shears* at the front there's a pin through your waggon body, through the turntable. But we had never bothered with pins in the bottom of the pin because if we'd have been going to unloaden it, the first thing we'd have done would have been unloadened off the back. So he pulls up, nobody said nowt to him, and he takes 'em off the front, and we wasn't there. But ' next thing we know, he comes. Foreman says, 'Have you got 'em loadened then?'

'No,' he said, 'we've had a bit of an accident,' and he took 'em off the front, and as soon as he'd got a ton off, what happened was it all went up in the air and all the lot went on the floor! 'Course, the pin had come straight out and as soon as all the taties come off, the waggon come straight down again and luckily the pin had gone straight back in the hole, and there they was on the floor.

Foreman says, 'Well, you'll have to pick 'em up.'

He says, 'I can't pick 'em up on me own!'

'Well, you wanted 'em on the waggon!'

Anyway, he let him sweat a bit and then we all went and loadened him up. But it'd have served the feller right if we'd have left him.

Now, at another time when they were coming for these five-ton loads, we were dealing with a lot of firms, and each tatie merchant brought his own bags. Some were all uniform bags with their name on, and some would just be dairy nut bags, any sort of bag. They'd all have eight stone, but some were full and some were only half full. So these lorries come and this driver said, 'I'm a bag short, Foreman.'

'Give him another bag.' So I put another bag on, and off he would go. So they were coming again, and Foreman says, 'There's only one man who never has enough taties, ' always wants another bag or another two bags.' So when we were riddling, there was two men shovelled into the riddle, one man turned the handle, another man picked the bad'ns off and another man was weighing them up, and then another man sewing them up and stacking 'em. And there was about yard to a ton, so every time you'd got about five or six ton, you then used to move your riddle down the pie. So he says, 'Now stack those bags in twenties, in pyramids! Five ton. Alright? Now, you've counted them, how many is there?'

'Hundred.' So he counted 'em.

'Hundred.'

'Now, come on,' he says, 'count these bags!' I'd count 'em: hundred. Everybody counted 'em on their own.

'Hundred.'

'You're all sure there's a hundred?'

'Aye.' So then we'd move down the pie, do some more for somebody else. Then this lorry come, and we loadened it up.

'I just want another bag, Foreman.'

'Oh, are you sure?'

'Yes.'

'Right,' he says. 'let's have 'em all off; and if I'm right you loaden
back yourself, but if you're right, we'll loaden for you. Take 'em off!'
So he handed 'em off and we stacked 'em on the floor and then we
counted 'em. 'Now then,' he says, 'just a hundred. Cheerio, we're
going for our dinners.' We left him. The feller had all these bags,
'cause when they come for five ton there was so many [bags] stood
upright, then they was doubled up laid flat on the top, and for every
ten bags you had on the bottom, you'd five on bags on the top, then
another five on top of them making a ton, and we all went for our
dinners and left him, and by, he would have a struggle, would the
feller. Because it's a struggle enough yourself to pick a hundred
weight up to shove it on a lorry. He never come that trick no more.
But he was a good feller was George Gibson, he knew his job.

Although the farms in the area were much alike in the way they
were organised, the characters of the different foremen, their experiences and preferences varied. So while George Gibson was very
much involved with the horses, and broke them in himself, Charlie
Buck left the breaking in to his waggoner. These foremen also had a
different approach to the horses. Whereas Charlie Buck went quietly
among the horses, George Gibson would make a lot of noise when
the young horses were first in the stable.

They'd all be tied up, and he'd get a metal bucket and he'd throw it
from one end of the stable to the other, and these horses used to
stand in the crib, looking straight down, and he used to say, 'They
won't be frightened of nothing,' and they weren't. You could have
thrown the bucket out ' the waggon, they'd heard it rattle that many
times. And he used to have a six-foot fork, and bang the beam at the
back, and then just let it fall in his hand, and it would just drop, and
it gained momentum and it hit 'em fair on their backs, and that made
'em jump. He was well out ' the way in case it kicked, and after he'd
done that a few dinnertimes, they didn't even move. He used to do
it every dinnertime, and by ' time he'd finished you could go in there
and you could throw a bucket up and down and they'd just all turn
round in the stall and watch the bucket go to and fro; never as much

as move. Now when you went to Pasture House, he wouldn't have 'em frightened like that. He used to be, 'Don't make a noise, you'll frighten them 'osses.' He had a different approach all together, and yet they were both men what had spent their lifetime among 'em, but he didn't believe in frightening them. But if somebody dropped a bucket outside, them 'osses of George Gibson wouldn't flinch.

No matter how capable a foreman was, there were still instances when things went wrong. One procedure that was open to accidents was pulling the waggons, or in this case a trailer, swing. While a vehicle would generally come gently to rest in a field, when on a hard surface the momentum would keep it in motion, and without a pole it was impossible to steer effectively. So it was not surprising that things did not go to plan when George Gibson had Ron leading sheaves straight from the field to the threshing machine.

We used to lead and thresh, especially barley. We had threshing machines set up in the yard; that's when all these German prisoners of war were all carrying caff and pulls. We was in the field and they were threshing. I remember once coming down with a load of sheaves and I couldn't stop it, 'cause I couldn't run it into the horses; and I run over the scales, then I run over the winding up barrow and then I hit the threshing machine. It was still crabbing and I had my 'osses in my hand, leading 'em, and it was still going. He saw he'd done wrong, like, but he never said nothing.

One of the advantages of the hiring system, particularly for a lad who hoped to become a foreman himself, was that he had the opportunity to work on a number of farms with different foremen. Although Ron was never hired at Pasture House, he did work with Charlie Buck occasionally, but being friendly with Charlie's son, Harry, Ron often used to visit after work.

Charlie was the real foreman; and everything was done with horses. And when I was at West Newton he showed me how to stack square-ended stacks with a box of matches on ' kitchen table. That's all I knew, and I went back to West Newton the next morning and stacked a square-ended stack.

In addition to explaining how to do various jobs, Charlie Buck also impressed upon Ron that he should take his responsibilities as a waggoner seriously.

He used to keep me right, did Charlie Buck. I used to go to their house a lot and he used to tell me what I should be doing, and he used to say, 'Now you've got to be Waggoner, you have to be strict and sociable. How can you drink beer with them on a Saturday night, and kick their arses on a Monday morning?' And he was quite right. He took life very serious, he never joked at work about anything, but when he wasn't at work, he was quite jolly. But, 'Never pull anybody's leg about their work,' he used to say, because they could take it very serious. We were very serious at work.

It was not just the farmers and foremen who took the work seriously. The ethos of always doing a good job permeated the whole workforce, so if anyone was failing to do their job properly, even those further down the pecking order took a dim view.

One year we got a foreman from ' top side of York, Strensall area, and it's a real tatie-growing area, that light land; and he come and he looked at these stacks, and he said, 'Who's stacked all these, then?'
I says, 'Foreman and one of the labourers.'
'Ooh, I can stack better than that,' and he certainly showed us a bit about doing tatie pies. He always used to use bottles when he was doing the tatie pies, it was all done and all spitted, all straight. It was only going to be for a matter of days* but it was all done correctly, and he could do these tatie pies.

But anyway, harvest comes, and he set off to stack; and me being Waggoner, I sort of had to tell him that you had to be careful where you put your stacks because you had to get your elevator to get your straw, and if you didn't stack it in the right order in the stackyard, you didn't get your straw in the right place. And the first three stacks were going to go down ' side of stable, but once he got above ' stable, the wind caught him and he says, 'Does it allus blow like this?'

And I says, 'It hasn't started yet.' We was only running one stack at the time, and we were down in fields loadening these trailers, and the next time I come up he'd started another one! He'd got as far as ' top of ' stable, sheeted that stack down and he'd started another one, and when we went in for our dinners his wife said to me, 'How long does it blow like this?'

---

* The potatoes might actually stay in the pies for months, it was just that the straw was only visible for a few days until the pies were soiled down.

I says, 'Well, I don't know, but it can blow most of harvest; it's only a breeze is this,' and then the next thing he'd started another stack! So we'd three stacks just as high as the top of the stable, and he says to me, 'How long's it going to blow?'

I says, 'No telling, but it's noted for blowing.' And any strangers who used to come from away used to say, 'How does it blow,' but this was what we knew. I mean, in the days with waggons, when you were loadening, very often you used to have to get somebody to sit on with a sheep hurdle when it was blowing, and that's why I always think we loadened as we did in Holderness. We could turn into the wind and the lad would sit on the front on the sheep hurdles and we'd put the back on. Then he'd put the hurdle on the back and we'd loaden the front. (See Appendix 2 on page 189.)

So he said to me, 'Will you come up on this stack with me? Let somebody else go down to field.' So he sends somebody else down, 'cause we were still pulling about with horses in the field. So I goes on the stack and I'm picking for him, and I says, 'If you don't get these topped up, once you start another row you can't get these elevators in,' 'cause you stack two rows of stacks so you can get the threshing machine down the middle.

So as we were going along he says, 'Can you stack?'

'No, I never had done.' I'd had a go, but I'd never actually stacked a stack.

So he says, 'Have a go at stacking a bit. I'll pick for you.'

So I was stacking away and he says, 'Are you alright? I've to go down ' field.' So he sent somebody up on stack with me; and he clears off! So I'm going along, and he never comes back. I thought, 'Well, it'll soon be time we're topping up;' we'd got half on it and I was on the top half, and he never comes back! So I start topping up and there's a special way you top up when you're outside to make 'em waterproof, and I finished this stack, and he says, 'Oh, you got that one finished then? Would you like to have a go on the next one?' So I had a go on the next one, and then they started to run two stacks. So he had to stack then, but he didn't like it. I think he only lasted either one year or two years, I can't remember. But he wasn't up to his job.

On the other side of the coin, those who took an interest in their work were respected by their elders. An example of this was when Ron was sent to stack bottles of straw one threshing day at Cliff Top Farm, Aldborough, where a man called Bob had been foreman for the Caleys for many years. Not only was Bob impressed that Ron knew how to stack, but also that he had stayed working on Caley farms; but he was equally scornful of the other lads messing about even though at the time they were not at work.

They were threshing, and the only place for these bottles was between two stacks, and it was a long narrow stack. I'd two lads on with me but I didn't know 'em at all, and I was stacking these bottles, and round comes Bob. 'Oh, you've stacked bottles before.'

'Aye, that's right.' So when it come dinnertime, there's a pub just opposite Cliff Top; we was all in this pub, we'd all had our dinners packed up, and along comes Bob.

'Now then, are you going to have a drink with me?'

So I says, 'Alright,' and we sat down.

He says, 'Where are you from then?'

'West Newton.'

'Oh. Where was you before that?'

'I was with Briggs at Marton.'

'Ooh, and before that?'

'I was with George Gibson at Flinton, and I've been at Pasture House with Charlie Buck.'

'Oh, you're a real Caley man!' He says, 'Look at these silly buggers!' This is all these lads performing about and playing darts and that. And these lads says to me, 'By hell, what have you done? You're well in with Bob.'

The lads' surprise at Ron being befriended by Bob was because although he also was a Caley man, like Ron, he was known to be difficult at times.

By all accounts he was a strange man to get on with, but he'd been with them most of his adult life, until wartime. I think he left them and went on munitions 'cause he could earn more money. Then after the war he came back. And they always tell the tale about Bob, when Gil Briggs sent his waggoner to Cliff Top for two carts; this was about five mile away. So when he gets to Cliff Top, he said he'd come for two carts and he said, 'Who's sent you?'

'Foreman.'

'Well, have you got a note from H Caley?'

'No.' So he had to ride back to West Newton, which would be two and a half miles, get a note from H Caley to come back to Bob, and when Bob read it, he says, 'Well, you can have one,' and sent him back again!

And on another occasion, Ron Hunter, who was a long-time worker for the Caley family, took the threshing machine up to Cliff Top, and Bob come out. This was after tea, 'cause when they finished threshing, they took it to the other farm to set up for ' morning. When he gets there Bob said what did he want, and he says he'd come to thresh.

'I didn't say I wanted to thresh, so I don't want one!'

Ron knew Bob of old, so took it down to Elm Tree Farm, which was in the village, left his machine there, got his bike and called in at West Newton, and they said, 'Oh, never mind. Have two or three days threshing at Elm Tree Farm!' So he had two or three days threshing there, and while he was threshing there, Bob comes along on his bike and said, 'When are you coming to thresh for me?'

Although behaviour such as this could make life a little difficult for the bosses and for the horselads, as long as the work was getting done properly, most people put up with minor eccentricities. For those lads and labourers who did not get on with the foreman, they always had the option of leaving at the end of the year, and if the boss was unhappy with a foreman he simply would not ask him to stop again at Martinmas. It was this flexibility which allowed the system to function, to the benefit of all concerned.

CHAPTER 13

Holderness and Beyond

THROUGH the 1950s Ron continued to work for Norman Caley, and largely because it suited him, he carried on being hired by the year, although by this time the hiring fairs had come to an end. There were some, including Ron, who still met up in the pubs near the statue of King Billy on Martinmas Tuesday, but these were now purely social occasions. At West Newton Ron was still in charge of the stable, but as the '50s drew to a close the horses were being used less and less.

> We'd now got nine wheel tractors and there was four crawlers, and we still had four or five horses and we did odd jobs with 'em, but we didn't use 'em that much. We still did use them, especially in wintertime, but they'd about come to end of their usefulness. 'Til they got these crawlers, we was the kings; they couldn't do half the thing without these horses, but once these crawlers came along, they could do the work of horses on very heavy wet land, whereas horses could harrow when tractors couldn't possibly. I've seen us harrowing with horses, and the tractor's gone right down to its axles within a foot of where we'd been, he's had to pull his pin out and leave the harrows where they were to get the tractor off.
>
> Even when we only had a few horses left and we were drilling wheat with tractors, we had a set of four horses abreast in gibb harrows and we always was last time over. With four, and with the speed that they used to go in them days, we could more or less keep up to 'em with a set of four gibb harrows. I think it was only a bit of pride, 'cause you was leaving no wheel marks, and then if we'd plenty of time I would cross it, and so it'd be nothing to have harrowed two hundred acres at the back end, what we'd sown with wheat.

For most of his time at West Newton Ron carried on living in the foreman's house and was paid by the year, although most of the other workers were now being employed in a more usual manner. But in the late 1950s Ron's conditions of employment changed, for the traditional reason that he was to marry. As had often been the case in the past when a farm lad met his future wife while she was working as a maid, Ron met Nancy when she came to help Mrs Caley after she had given birth to a baby. Nancy's brother, Eddie, also worked at Old Farm and one day after work, he and Ron went to his family's house at Norwood, and the three of them decided to go to see a film together. After the film Ron and Nancy walked back to West Newton, starting a romance which led them to marry in 1959. Once they were married, they moved into the Old Lodge of Burton Constable Hall, a few hundred yards from the farm where Ron was first evacuated.

The following year was the last harvest that there were stacks of corn at West Newton, ending another chapter in the story of traditional agriculture. The new developments on the farm and in Ron's private life ushered in a change which would take Ron away from his native Holderness. Ron had always had an interest in other places, relishing his trips out of the area, and now that he was married, he started to think about finding a job elsewhere. This thought-process may have been spurred on by a chance conversation with a visiting vet.

We got a vet come to do something with some bullocks, and he says, 'How much money are you earning here?'

'Oh,' I says, 'same as t'others.'

He said, 'You'd earn a lot more if you got away.' Then it come 'luance time and he set off to eat his 'luance, and he says, 'Aren't you having any?'

'No,' I said, 'we don't stop for 'luance, the only time we stop for 'luance is when they provide it.'

He says, 'You'd do a lot better down south.' So it set me thinking. So every Sunday night I got ' Farmers' Weekly, and I saw this job on the Cotswolds for a shepherd, and Norman Caley come out one morning, this was nearly Martinmas, and he said, 'How many more bullocks are you going to put in that yard?'

I said, 'I don't know, and what's more, I don't care; because I'm leaving you on Saturday,' and he turned on his heel, never said a word. He come back three days later. He said, 'What did you say to me?'

I says, 'I'm leaving you.'

'When?'

'On Saturday.'

He says, 'Who's going to cut all them pigs? There's about two hundred little pigs want cutting and there's two hundred calves want cutting, and you've all them *cock bods* to *caponise*. He said, 'You'll give me that feller's telephone number. I'll soon ring him up!'

I said, 'Well, I'm not stopping with yer, but I'll put it off for a month,' so I rung this feller up and said I wouldn't be going for a month, and I cut all these pigs and I cut all these calves, did this and did that, and be that, it was Martinmas week. And I left and I went on the Cotswolds.

The new job in Gloucestershire turned out to be a big disappointment. Even on the first day when he was shown the sheep in the different fields, and was told which order to go round them, instead of finishing back at the farmstead the tour ended up with him being at the very far end of the farm! Later when there was some fencing to be mended, he was reminded to take a hammer with him! For a man used to being left to get on with the job and being in charge of other men, this was not the treatment he was used to, so he soon started looking for another job.

A year later Ron applied for a job as a farm bailiff in Sussex, and at the interview was asked for a reference.

He asked me if I'd any references and I said, 'Well, I've never had a reference in me life,' I didn't need one. Well, he couldn't understand that, so he said could I give him a telephone number. So I gave him what I thought was Norman Caley's telephone number. I'd never ever rung Norman Caley up, I just recalled there always was a number on a lorry, and I got a telegram back. 'Not known. Would you ring me at such and such a time and reverse the charges?'

So I rung him up and he says, 'I've rung up, and they put me in touch with a Peter Caley and a George Caley, but no Norman Caley: all these other Caleys, but none of them's him.'

'Oh,' I says, 'there is a Norman Caley, but I must have got his number wrong.' So he says, 'I'll try directory enquiries,' which he did, and the minute he'd got through I got a telegram, 'The job is yours. Ring me at such and such a time,' to arrange for me to go.

It was during the cold winter of 1963 that Ron, Nancy and their two small children, Anne and Mark, packed up to go to Sussex. The new job was for a newspaper owner, F J Parsons, whose farm covered 400 acres, in amongst another thousand acres of woodland. Mr Parson's main interest in the farm was the herd of Sussex cattle, which are a horned breed native to that area, but the aim of Mr Parsons' breeding program was to produce polled Sussex cattle, starting by using a polled red Angus bull to introduce the polled gene into the herd, and then breed back to get an animal of a Sussex type.

Ron's job required him to organise all the work on the farm, and F J Parsons was more than happy to leave him to get on with it. This part of the job was straightforward for Ron, but at first he did not appreciate that being a farm bailiff also involved buying and selling, a side of farming which Ron knew little about. It was only when Mr Parsons asked whether it was about time that some lambs were sent to market that Ron realised he needed to get to grips with this aspect of the job. Nonetheless, he soon learnt his way round the markets, and on one occasion was characteristically unafraid to intervene when things were not as they ought to be.

> I went in this market, and these fellers were bidding, and the auction-eer stood and he said, 'Going going . . .' and I shouts, ''od 'ard a bit. Six pounds!' So these other lot had a bid, and so I had another bid and then I says, 'Right, now you can have them,' and they'd got a little ring going. All these dealers were stood together and they were going to bid five pound and then would share out among theirselves. By hell, they did look at me!

Back on the farm there were also other problems to resolve, as Ron's predecessor had let things slide and the farm was poorly run, which was quite a shock to someone brought up in the Yorkshire tradition. However, it did mean that Ron could improve the farm and bring it back into profit. Although Mr Parsons did not interfere with Ron in his running of the farm, the reverse was not true, Ron's forthright manner being a source of intrigue, and quite a surprise to Mr Parsons, who was not used to his employees talking to him in such a manner.

> One time he said, 'Will you send the men up to dig the garden?' These fellers told me what would happen, so I says, 'How many gardeners have you got?'
> 'Well,' he says, 'there's three gardeners, a handyman and a groom.'

'And,' I says, 'you want the farm men to go and dig your garden?'
'Yes.'

'Well,' I said, 'if when I come up here and I seen all these garden-ers fully employed, I will come and dig the garden meself!'

'Oh no,' he says, 'I don't want you to come and dig it.'

'No,' I says, 'I didn't say I would. What I said was when I see all this lot fully employed, I'll come dig it meself!'

'Well,' he said, 'aren't they employed?'

I says, 'I don't know. Who's in charge of them?'

'I am.'

I says, 'There's your answer!' Well, they never went. I says, 'You've more bloody fellers up here! Dig the garden! They want to paint your house!' Well, they loved me did the gardeners!

Another time he says, 'I'm going to Goodwood.' He was always going racing.

'Aye,' I said, 'I know what I'd do with you all.'

'What's that?'

I said, 'When you all got inside I would shut all the gates, then I would let you all out and give you all a pick and shovel, and say we're going to build a road!' Oh, he used to love it; he used to say nobody had ever spoken to him in his life like that, and 'course, he used to tell all these tales at these dinner parties, and the chauffeur used to hear 'em all, and he used to come back and tell me.

The years spent in Sussex while the children were small were a settled time for Ron's family, and Ron's work brought him in contact with people from different parts of the world involved with cattle breeding. None of his work involved horses, but soon after arriving in Sussex he was asked to help a man with his three Shires, which led to Ron spending his spare time helping different people with their horses, and taking them to shows and other events all over the south of England.

After F J Parsons died, Ron did various jobs with horses, running horse-drawn omnibuses in London for two seasons, and looking after a herd of Lippizaners for a Hungarian bullion dealer. Then in the early 1990s he moved to Cambridgeshire, working for a time as the horseman at the National Trust's Wimpole Home Farm. On retiring, Ron continued to be just as active, going round the show circuit, training and showing other people's horses, and giving talks about his early years as a horselad. In 2004, he and Nancy returned to

Holderness, living only a few miles from where they spent their formative years, among people sharing a similar culture and history.

> Only the other month; well, it was Mart'mas week, I was in Hedon, and I met a feller and he shouts, 'Are you stopping again?' and I says, 'It bloody looks like I'll 'a' to do!' And we was stood in Hedon marketplace talking.
>
> 'Here's a bugger, look,' and he shouts, 'are you stopping again, George?' And he come across and they were talking, and fetching me into ' conversation, and all three on us had been at Caleys' at some time.

This last incident was recorded during my last visit to see Ron, when he continued to fill in the details of his life as a horselad, and relate new anecdotes, full of detail, full of life and passion, often with a mischievous twinkle in his eye. During the previous few weeks he had started to have trouble walking any distance, caused by restriction in the blood vessels near his heart. He died following a heart operation in April 2008.

AFTERWORD

IN many ways, this book would have better described the life of an East Yorkshire horselad if it had been written fifty years earlier; if it had been written about someone of Charlie Buck's generation who came of age at the end of the nineteenth century when the horse was supreme. Even someone a little younger, who was working before the introduction of the agricultural wages board and the desperate years of the 1930s, might have given a more rounded picture of the horselads' world. But we are lucky to have this story at all, because not only was Ron born just in time to catch the last days of the farm horse and the hiring system, but also, he related his story as an old man. In the intervening years the only record of the events of his early life was in his memory: a precarious connection to the past perhaps, but in Ron's case a rich stream of detailed knowledge and anecdote.

Of course, since Ron was a horselad, farming has changed dramatically. The process of mechanisation which ousted the horses has simply continued, the earlier tractors being replaced by machines of increasing size and complexity. In 2008, for instance, Harry Buck told me of a local farmer with a combine which can harvest forty-seven tons in an hour. Especially for someone who grew up in the era of threshing machines, that speed is fantastic. As Harry pointed out, it would have taken twelve men three days just to thresh that amount of grain, plus all the work at harvest time.

Although this efficiency is remarkable, it has come at a price, the most obvious being the lack of people involved in farming, so that most country dwellers now have little connection with the land. There are also costs which are less apparent: the price of agricultural subsidies and the straightjacket they impose on farming systems, the loss of wildlife and organic matter in the soil, and the costs water companies incur in removing pesticide residues from water.

Even if we believe that these are costs worth paying, the notion that modern farming is efficient is only true for certain criteria: although it is certainly efficient in terms of money and labour, it is extremely inefficient in its use of energy. From powering tractors and drying grain, to making fertiliser and transporting food, agriculture uses a vast amount of fossil fuel; and in the process, that fossil energy is converted into a far smaller amount of energy in the form of food. If we were to do the same thing with money, rather than energy, we would have a global business that always runs at a loss, a system that is bankrupt. It is a situation that cannot last, not only because oil reserves are insufficient to meet demand, so we will inevitably have to use less, but also because the change in the climate tells us we need to do so quickly.

In the changes that farming is just beginning to make, we will need to develop systems which are net producers, rather than consumers, of energy. In that search, there will be lessons to learn from the past. As new farming systems are developed, horses could well play a significant role, but whether that happens will depend on our priorities and the available technology. We may perhaps have effective tractors running on solar power, through the use of photovoltaic panels, and wind power; or we could employ draught animals, using power from the sun to grow grass and oats for them to eat. More likely we will use a mixture of different methods. It is within this context that stories such as Ron's may, in the end, have their greatest value.

APPENDICES

Appendix 1 – Rotations

Before the introduction of potatoes during the war, farms in Holderness followed rotations that were fixed, so that the old labourers Ron worked with would work out when an event happened by knowing which field the clover seeds were in that particular year. Although these rotations were no longer being followed when Ron was at work, he thought that High and Low Fossom Farms followed a seven-year rotation, whereas West Newton used a five-course rotation. This five-year rotation was followed at Old Farm and probably at Carr Farm. The last year that Carr Farm had any bare fallowed fields was 1947.

The likely course of the rotations went as follows:

<u>Year one</u> – barley or oats, undersown with clover. (Sown at 10 lbs white clover and 5 lbs red clover to the acre.)

<u>Year two</u> – seeds, grazed by sheep, and ploughed up in autumn

<u>Year three</u> – wheat

<u>Year four</u> – bare fallow. The peas and mangels were also in this year of the rotation.

<u>Year five</u> – wheat

Appendix 2 – Loading sheaves

The method used to load sheaves onto a vehicle in Yorkshire differs from the method used in the rest of the country. In Yorkshire there are two different ways of putting on a course (or layer) of sheaves, the square course and the shipping course. When loading on a rulley, the first course is a square course.

In a square course the first sheaves are placed transversely with the heads towards the centre, and the butt end overhanging the vehicle by a few inches. Starting at the back the sheaves are placed side by side until they are near the front of the vehicle. (See Key 1.) Then the last three or four sheaves are put on, starting at the front, so that the sheaves on the corners always have some of their straws trapped by

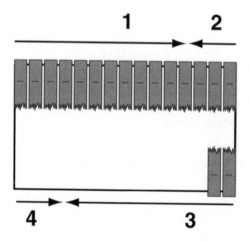

the subsequent sheaves. (See Key 2.) The other side is done in a similar manner. (See Keys 3 and 4.) Two or three sheaves are then laid between the existing rows at the back, with the heads facing inwards (see Key 5), and then sheaves are laid in a herringbone

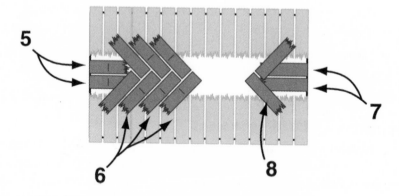

pattern down the centre of the rulley, overlapping the heads of the outside rows. (See Key 6.) The centre at the front is finished in the same way as the back, so completing the course. (See Keys 7 and 8.) Ron would lay another square course, and then use shipping courses all the way to the top of the load, whereas some people put a shipping course between the two square courses.

In a shipping course the sheaves are placed longitudinally. Starting at a corner, say the back corner, the first row is placed with the heads facing forwards, the loader moving a few straws from the first sheaf so that these are trapped by the next sheaves. (See Key 9.) When the middle is reached, the other side is put on in the same way, though often a single sheaf, known as a lifter, is laid transversely (see Key 10) before the corner sheaves are put down (see Key 11), in order to keep

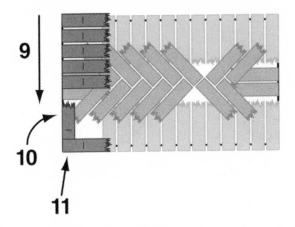

the corner up. The next row of sheaves, which are placed the other way round with the heads facing upwards, partially covers the first row. (See Keys 12 and 13.) The loader continues to lay rows of sheaves partially covering the previous row, always starting at the outsides, until he nears the front of the vehicle. The front is started in

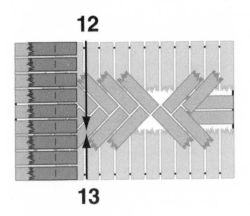

the same way as the back. (See Keys 14 and 15.) The final row of sheaves in a course has the butt ends resting on sheaves from the same course, so binding the back and front of the load. (See Key 16.) The next shipping course starts at this end of the vehicle, so the loader works from front to back with one course, and then from the back to the front with the next course.

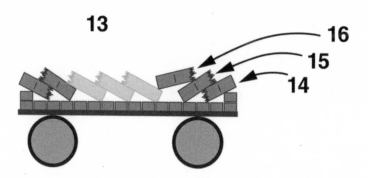

Traditionally the loads became narrower with each course so the top course was only one or two sheaves wide. This meant the load was more stable than if the sides had been kept straight. When loading a waggon rather than a rulley, the body was first filled with a shipping course to bring it up to the level of the shelvings, and then it was loaded like a rulley.

In other parts of the country the whole load consisted of square courses. This creates a more stable load, but it is harder to fork sheaves over the butt ends of the sheaves in a square course, than the sides of the sheaves in a shipping course. Also the man on the load has to move more when doing a square course, so each load – and there-fore the whole harvest – takes longer.

When working in the field it is easiest when the load is downwind from the forker, but when it is very windy the butts of the sheaves facing the wind can lift up and flip over. When putting on a shipping course, it is only those sheaves at the front of the load and at the back which have the butts facing outwards, and are therefore in danger of lifting, and as soon as the next row is put on, this danger is much reduced. This is why Ron thought that the loads were built in this way in Yorkshire, as even when it was extremely windy, it was only the front or back which would need a hurdle to keep the sheaves down, rather than the whole length of a waggon.

Appendix 3 – Loading sacks

The pole waggons were designed to carry a load of three tons, over any terrain, but on threshing days, when they were merely used to hold sacks of corn waiting for the lorry to come, considerably more was put on. In the days of taking the corn to the mills, there was a special way of stacking the sacks. Ron was once shown how to do this by George Gibson, when they were moving corn from one farm to another. After the corn had been wound up to shoulder height on the winding up barrow, the sacks were carried up a step ladder and dropped into position on the waggon. All the sacks were stacked lengthways, in three sections, each called a dess. In the waggon body two sacks were placed side by side at the front, then two at the back and two in the middle. The next course consisted of three sacks in each section, one being placed on each shelving and one in the middle. The sacks in the front dess had the mouths facing backwards, while the others faced forward. This allowed the corners which showed at the front and back of the waggon to be pulled up to make a smart impression. The next course had two sacks placed on edge, the following course also having two sacks, and the final course consisted of a single sack. Each dess therefore consisted of ten sacks, making a waggon load of thirty sacks in all.

When corn was sold by volume, it was sold for so much money a quarter. A quarter was made up of two sacks, each of which was four bushels, a bushel being eight gallons. So a load of thirty sacks was fifteen quarters. Because of the different densities of the various types of grain, a quarter of the different grains varied considerably in weight. With barley, each sack weighed sixteen stones, so a waggon load of thirty sacks weighed three tons. In the second photograph section, the photograph of corn being taken on the road with six sacks in each dess above the shelvings, and two out of sight in the body of the waggon, making a load of twenty-four sacks, suggests that a heavier load of peas or beans was being transported.

GLOSSARY

'a' – have

airbreed – strouter, metal bracket to support the shelvings on a waggon

allus – always

an' al [and all] – as well

artificial – artificial [chemical] fertiliser

artificial drill – machine for spreading fertiliser

arve – see orve

arving out – splitting a piece of unploughed ground between two ridges or lands, when the horses turn left at the headlands

bacca – tobacco

back band – leather strap which holds up the trace chains

back door lad – lad employed to work round the farm yard and house doing odd jobs

bag – sack

bearing rein – strictly speaking a hame rein. A rein which goes over the top of a hame to stop a horse putting his head down

belly band – leather strap which keeps the traces, or shafts, from rising too high and so choking the horse

bit ring – ring at either end of the bit

blather – mud

blinders – bridle with blinkers

breast pole – or neck yoke – the bar at the front of a pole which attaches to the collar to keep the pole up

breech band – see britchin

britchin – strap around the back of a horse, and those used to keep it up, involved with stopping or backing a vehicle

bodkin – method of yoking three horses, typically when ploughing, where one horse walks on the land and the other two are in the furrow

box – loose box

bottle – truss, low density bale approx 4 ft × 2 ft × 10 in.

bowline – knot used when tying cords on bit rings

bullocky – man employed to look after the bullocks

caff – see chaff

caff hole – place in the stable where chaff was stored

caff sheet – a sheet, tarpaulin, or hessian used for carrying chaff

caponise – castrate [poultry]

cart – two-wheeled, horse-drawn vehicle with shafts

cart saddle – sits where a riding saddle goes, to hold up a pair of shafts

chaff – the casing of a grain used to provide bulk and fibre when feeding horses

check – single line or rein used to control a horse

check line – see check

check horse – horse that was driven with a check rein

clever feller – smart-arse

clog – lump of wood, often spherical, with a hole through. Used to provide weight when tied on a halter shank

coconut band – string made from coconut fibre

cock bod – cockerel

cords – ropes, usually made of hemp, between the horse's bit and the horseman's hands used to control the horse

cowler – tool like a big hoe used for scraping muck

coulter – knife on a plough used to cut the vertical part of a furrow slice

couping – type of stack

coupling band – light rope fastened between the bit rings of two or more horses, so they turn together

course – layer, usually of sheaves or bottles

crooks – gate hooks

crossed helters – the halter shanks of typically two horses, each tied back to the hames or dropped trace links of the other horse

curb bit – bit which has a curb chain which goes under the horse's jaw dess – a lump of hay cut from a loose hay stack (see page 131)

drag – hand tool with two prongs at right angles to the handle, used for picking up roots such as mangels

draught – a vehicle

draughts – the eveners which equalise the load between a team of horses and the swingletrees which keep the trace chains away from the horse's sides

dyke – ditch

etch – tail board on a cart or waggon

fallow – land which is not growing a crop in order to reduce weed numbers by cultivation

false line – special type of rein which clips into another horse's trace

farside – horse on the right-hand side of a team when viewed from behind them

fast – tight, stuck

fastening penny – see fest

feller – fellow

fest – sum of money given by a farmer to a hired lad, which sealed the agreement for the year's work

fetching – cutting a crop down one side of the field, usually because the crop is laid

Fiver – the lad's position in the hierarchy below the fourth lad

Foreman – employee in charge of the horselads and labourers

form – bench

forker – usually the man in the field who forks the sheaves onto a load

for – furrow

fost oss [first horse] – trace horse, the horse used in trace gears, yoked in front of another

Fowth Lad [fourth lad] – horselad whose position in the hierarchy was below the waggoner and the third lad

gap – to remove plants, typically mangels or swedes, out of a row to give the others enough space to grow

garings – triangle-shaped pieces of land left when the rest of a field has been worked in parallel pieces. Also in this case, where the furrow has been left crooked and uneven

geared – harnessed

gearing or gears – harness

gee back – command for a horse to go to the right

gibb harrows – chisel harrows, where the bottom of each tine is angled forward and sharpened like a chisel

gormers – harvest ladders

grass reaper – mower, in this case a sickle bar mower

green topping – lifting potatoes before the skins have set for new potatoes

hame – parts of the harness which fit to either side of the collar, and to which the traces are attached

hame hook – the part of the hame on which the trace chains hook

hame ring – a ring on the hame, through which the cords or reins pass

hap – to cover with straw

headland – the piece of ground at the end of the field used to turn the horses

heck – rack, feed rack for hay

heck meat – hay, or sometimes straw [heck meaning rack, meat meaning food; therefore food in a rack]

helter – halter

helter shank – rope attached to the halter

hicking stick – stick used by two people under a sack to assist lifting it

jack straw – large fork with two-foot-long tines used for moving loose straw

jibbing – refusing to go forward, usually used about a horse

laid – corn that had been flattened by the weather

laid – harrows which had had new points hammer welded onto the tines

land – the higher portion of a field between two water furrows

lead – to take a load somewhere

leader – horse[s] in front of another

leader – person who led a horse

leading – the process of taking a load or loads somewhere

Least Lad – the lad at the bottom of the hierarchy

lees – lies

ley – temporary crop of grass and/or clover

loaden – to load

long tom – pitch fork with a six-foot shaft

looking – to look for and then to hoe out weeds

louse out, loose out – to unhitch

'luance [allowance] – a drink and a small amount of food taken mid-morning or mid-afternoon

maister – master

master tree – the evener used to equalise two horses

meeter strap – leather strap connecting the cart saddle to the collar, or a britchin to the collar

monkey – wooden platform with two legs secured to a stack for the picker to stand on

muck hill – muck heap

muck road – a pair of furrows ploughed in a fallow field for the horses and the waggon wheels to go down

name for nowt – to be reprimanded

nearside – the horse on the left side of a team when viewed from the back

now then [sometimes pronounced 'noo then'] – greeting, like hello

nowt – nothing

offside – farside, the horse on the right-hand side of a team when viewed from the back

'od 'ard [hold hard] – wait, stop

over end – straight up

owt – anything

pie – clamp, a heap of root crops covered with straw and soil to protect the roots from frost over winter

pike – a stack of corn with a circular base

pole chain – one of a pair of chains attached to the front of the pole on a waggon, which hook onto the hame hook

pot picker – a T-shaped tool consisting of a handle and a metal rod which is pushed into the soil to locate land drains

pulls – broken pieces of straw and leaf separated from the straw and grain in a threshing machine

rakings – hay or corn raked up in the field after the haycocks or sheaves have been stacked

reapings – the cuttings of hay, brambles and other weeds cut from the side of a dyke

riddle – machine consisting of various sizes of mesh used to sort potatoes

rigg – Yorkshire version of ridge. In this case it is the first piece of a ploughed land, when the first two proper furrows are thrown together

rulley – four-wheeled vehicle without sides, either with shafts or a pole

sail – wooden reel on a binder which pushes the top of the standing crop backwards as the knife cuts the crop

scotch bob – way of plaiting up a horse's tail

scruffle – to hoe with a horse [or tractor] hoe

scruffler – single-row cultivator or hoe

scuttle – metal container with two handles

seeds – clover ley

sew – sowed

shaft horse – horse that is working between the shafts

shak – shake

sharps – by-product of wheat, after it has been milled

shav – sheaf

sheaf – bundle of corn tied by hand or by a binder

shears – the fore part of a four-wheeled vehicle which turns

sheet – a tarpaulin of canvas or hessian

shelving – lade, the 'shelves' sticking out from the side of a waggon over the wheels

shim – horse hoe for more than one row

shot – when a sack of corn has been emptied out on the floor

side delivery – side delivery rake, hay turner with a reel running at an angle to the direction of travel

side string – rope rein which goes to the horse's bit

skelbase – partition in a stable between standings

skeef – disc, disc coulter

slash knife – bill hook

slag – basic slag, a by-product of steel-making used as a fertiliser to provide phosphate

sneck – door latch

sock – the point or share of a plough, the part which cuts the bottom of the slice and starts to lift and turn the furrow

standing – stall, section of a stable with space for one horse to stand [some were wide enough for two horses]

stook – a number of sheaves, stood up in pairs to allow the grain to ripen and dry

stooker – someone who is putting sheaves into the stook

stop – stay

swagger ball/knot – knot made in the end of a halter shank to prevent it passing through a clog, ring or chain link

swing – ploughing without wheels

swing – pulling a vehicle or implement without shafts or a pole

swing chisels – wide chisel harrows made in one section

swingletree – wooden or metal bar to which a horse's traces are attached

tandem – one horse hitched in front of another

tank – trough

tatie – potato

team – to unload

three-horse baulk – largest evener when using three horses

thresh – to separate the grain from the straw

Thod Lad [third lad] – on larger farms this could mean the third lad's lad, but here it means the lad under the waggoner

Thoddy – the lad under the waggoner

tilth – the state of the soil when it is broken down into small lumps

Tommy Owt – hired lad who would work with the sheep, cattle, or horses as required

trace – chain between the hame and the swingletree

truss – low density bale

tumbril – wooden feeding trough

tupped – when the ewes had been put in lamb by the *tup*, or ram

unicorn – three-horse hitch where the single horse at the front is central to the vehicle or implement

Wag – shortening of Waggoner

waggin – waggon

Waggoner – horselad in charge of the stable and the other horselads

waggon – four-wheeled vehicle with sides

wheeler – in a team of more than one row of horses, the horses nearest the vehicle

while – until

whoa – command for horse to stop [also pronounced 'whee, woo, whey']

wuzzel [mangel, mangold or mangel-wurzel] – swollen root related to sugar beet, used for animal feed

yer – you

yon – that, far: as in 'yon feller'– that fellow; 'yon end' – that end, the far end

BIBLIOGRAPHY

Primary sources

Most of this book is based on about thirty hours of recordings I made of Ron Creasey in 2006 and 2007. Audio copies, transcripts and copies of the photographs have been deposited at the Hull History Centre and at the Museum of English Rural Life in Reading.

Neil Lanham's DVDs of Ron Creasey, made in 2007, were another great source of information. These DVDs contain many stories and anecdotes about Ron Creasey's life that are not in this book. For more information contact Neil Lanham, Helions Bumpstead Gramophone Company, The Swallows, The Street, Botesdale, Diss, Suffolk, IP22 1BP. Email: **info@traditionsofsuffolk.com**

I also drew from my recordings of conversations, made in 2008, with Norman Caley, Tim Caley, and Harry Buck, as well as from numerous conversations with Geoff Morton.

Secondary sources and background

Beach, Bob, 'Yorkshire Waggons', *Heavy Horse World* magazine, Spring 1995.

Day, Herbert, *Horse Farming through the Seasons*, Hutton Press Ltd, 1991.

Day, Herbert, *Horses on the Farm*, Hutton Press Ltd, 1981.

Day, Herbert, *My Life with Horses*, Hutton Press Ltd, 1983.

Day, Herbert, *When Horses were Supreme*, Hutton Press Ltd, 1985.

Dee, Dennis, *Holderness Times and Arab Horses*, Privately published, 1997.

Caunce, Stephen, *Amongst Farm Horses – the Horselads of East Yorkshire*, Alan Sutton Publishing, 1991.

Chivers, Keith, *The Shire Horse: A History of the Breed, the Society and the Men*, J A Allen & Co Ltd, 1976.

Creasey, Ron, 'Yorkshire Check Reins', *Heavy Horse World* magazine, Spring 1990.

Hartley, Marie and Joan Ingleby, *Life in the Moorlands of North-East Yorkshire*, J M Dent and Sons Ltd, 1972.

Keegan, Terry, *The Heavy Horse, Its Harness and Harness Decoration*, Pelham Books Ltd, 1974.

Kitchen, Fred, *Brother to the Ox: The Autobiography of a Farm Labourer*, 1942.

Powell, Bob, 'Fenland Single Line Driving', *Heavy Horse World* magazine, Winter 1989.

Reffold, Harry, *Pie for Breakfast: Reminiscences of a Young Farmhand*, Hutton Press Ltd, 1984.

ABOUT THE AUTHOR

William Castle was brought up on a farm in North Yorkshire and has always had an interest in older methods of farming. In the early 1990s he worked with Geoff Morton, who continued to use horses on his farm near Holme-on-Spalding-Moor in the East Riding. Since moving to Shropshire, William uses his Percheron mare to work his small-holding in between making a living as a violin maker. He also writes regularly for *Heavy Horse World* magazine.

Other Books and DVDs from Old Pond Publishing

Beauty, Bonny, Daisy, Violet, Grace and Geoffrey Morton
Frank Cvitanovich

Horseman Geoff Morton chooses a stallion at the Shire Horse Show and puts him to a mare. The birth of a foal is movingly captured and other scenes show the horses at work on the farm and in the woodland. This Thames Television documentary won a BAFTA award and the Prix Italia. DVD

The Working Horse Manual 2nd Edition
Edited by Diana Zeuner

This guide to how heavy horses are kept and used today gives concise accounts of all aspects of heavy horses, including breeds, husbandry, farriery, work, multiple hitches, equipment, showing, road driving and vehicles. Paperback

Harnessed to the Plough
Roger & Cheryl Clark with Paul Heiney

A year on the Clarks' farm run with Percherons and Suffolks. They show Paul Heiney how to plough with horses, cultivate, make hay, harvest, build a stack and much more in an absorbing programme that recaptures the era when farms were focused on horses and horsemen. DVD

First Steps to the Furrow
Roger & Cheryl Clark with Paul Heiney

Paul Heiney watches the progress of Taffy, a three-year-old Suffolk Punch, as he is schooled for work on the farm. Cheryl shows the persistence and devotion required to train a horse for work in this engaging programme. DVD

Know Your Horses
Jack Byard

Jack's selection of 46 breeds shows the immense diversity within the species. He includes working horses, riding horses and ponies. Nearly all these breeds can be seen in Britain today and Jack includes fascinating details about their role in human history. Paperback

In a Long Day
David Kindred and Roger Smith

These 200 captioned photographs from the Titshall brothers show farm work and village life in Suffolk between 1925 and 1935. The selection includes working horses and their horsemen, steam-powered threshing and rural trades and transport. Paperback

The Horseman's Tale
Ray Hubbard

Former head horseman Ray Hubbard recalls his life working with Suffolk Punches in a beautifully filmed programme that includes the Hollesley Bay horses, archive footage and horses working in a Welsh woodland enterprise. DVD

Free complete catalogue:

Old Pond Publishing Ltd, Dencora Business Centre,
36 White House Road, Ipswich IP1 5LT, United Kingdom
Secure online ordering: **www.oldpond.com**   *Phone: 01473 238200*